Phaselock Loops for DC Motor Speed Control

Phaselock Loops for DC Motor Speed Control

Dana F. Geiger

A WILEY-INTERSCIENCE PUBLICATION

JOHN WILEY & SONS   New York • Chichester • Brisbane • Toronto

Library of Congress Cataloging in Publication Data:
Geiger, Dana F
  Phaselock loops for DC motor speed control.

  "A Wiley-Interscience publication."
  Includes index.
  1. Data tape drives–Automatic control.
2. Electric motors, Direct current–Automatic
control. 3. Phased-lock loops. I. Title

TK7887.55.G44   621.46'2   80-295 78
ISBN 0-471-08548-0   AACR 1

Printed in the United States of America

10 9 8 7 6 5 4 3 2 1

Preface

In the early 1960s the PMI Motors Division of Kollmorgen Corporation was manufacturing and marketing the first commercially available motor control phase locked loop (PLL) systems under the trade name "The S-1 Servo." Although it was usually possible to obtain excellent performance by using a phase frequency detector, servo compensation for a new system was often troublesome and required extensive trial and error techniques. Because of the engineering time required to build systems to meet varying customer performance specifications, PMI began to curtail marketing of S-1 Servo Systems in the late 1960s and was essentially out of the business by 1970.

Interest in the phaselocked servo did not cease, however, and PMI was often asked to quote systems or provide aid in applying them. This led in 1977 to reconsideration of the PLL motor control system. Progress occurred in the following steps:

1 Since the servo was considered to be a "moving position servo," which often uses "rate damping" to help stabilize the system, a rate damping scheme was attempted. This set the stage for a further improvement.
2 The up-down counter with D/A converter was used in place of the phase frequency detector. This yielded a frequency locked loop.
3 The analysis concept was changed from the "moving position servo" to the "velocity servo with integration." The ω (shaft speed in radians per second) was adopted as the independent variable.
4 The analysis was done in general terms (suitable for solution by computer) instead of being a graphical treatment of each case.

5 A four-quadrant velocity detector and op amp phase control circuit were added.

Among the contributors to these developments was Frank Bock who was the first to suggest the use of an up-down counter to replace the phase detector. Howie Bernstein of Kulite Semiconductor provided invaluable guidance in analytical methods. This led to the computer aided design approach instead of the graphical analysis for each case.

Other contributors were Frank Arnold of PMI Motors, who insisted on having a PLL, rather than just a frequency locked loop; Mark Stern, who always has a wealth of good ideas and suggestions; and Hans Waagen of Motion Control Inc., San Jose, California (formerly Chief Applications Engineer at PMI), who did the initial editing, provided many suggestions for improvement and gave me confidence that it "could be done." Connie Owens of the Engineering Department at PMI did most of the typing of the manuscript. Her enthusiasm and cooperative nature helped immeasurably.

DANA F. GEIGER

Port Washington, New York
June 1981

Contents

Phaselock Loops for DC Motor Speed Control

1

Introduction to Motor Control Phaselock Loops

Speed control of DC motors has been an active area of motion control since the invention of the first DC machine. Mechanical governors and then DC tachometer feedback (with all its variations) have dominated speed control methods. These conventional velocity servos often require that an error exist between commanded velocity and actual velocity. Although this error can be reduced by increasing servo gain, it cannot be reduced to zero. At some level of gain the servo becomes unstable.

The problem of finite error can be overcome by introducing a pure integration into the forward path of the servo loop. The pure integrator produces "infinite" DC gain and low "AC" gain simultaneously (see Appendix B). This avoids the oscillation problem (by having low AC gain) and allows for a servo with zero steady state velocity error (due to infinite DC gain).

Seemingly all the problems have been solved. This is not the case, however. Generator tachometers are relatively poor transducers of velocity. They are subject to all the vagaries of an analog component. Variations in air gap, temperature, and brush wear produce changes in the gradient (i.e., back EMF constant) of the tachometer. Since the servo system depends entirely on "knowledge" of the speed, the relatively poor capabilities of the analog tachometer are a fundamental limitation that cannot be overcome by sophistication in the electronics. It is for this reason that the phaselock loop motor control system is considered.

The phaselock loop (PLL) motor speed control system produces state-of-the-art speed control. The speed accuracy over an integral number of revolutions is virtually perfect! There is zero error between the commanded speed and actual speed. The speed accuracy over fractions of a revolution depends on the quality of the optical tachometer. The basic motor control phaselock servo is shown in Fig. 1-1. The DC motor has an optical disc mounted on its shaft. The disc has N radial lines on its surface. An LED (light emitting diode) or an incandescent lamp is on one side of the disc and a photosensor and grating are fixed on the other side of the disc. As the disc rotates, light is alternately transmitted or blocked by the disc/grating combination. The resulting output waveform of the phototransistor is a quasi sine wave, which is usually amplified and "squared." The frequency of this waveform (f_{tach}) is directly proportional to shaft speed [$f_{tach} = (N \times rpm)/60$ where $N =$ disc density].

The function of the phaselock electronics is to detect differences between f_{tach} and f_{ref} and drive the motor so that a cycle-for-cycle correspondence between the tachometer and reference frequency exists. Motor speed is thereby precisely synchronized with the reference oscillator. Instantaneous accuracy depends on the quality and alignment of the optics. (See Chapter 2.)

2

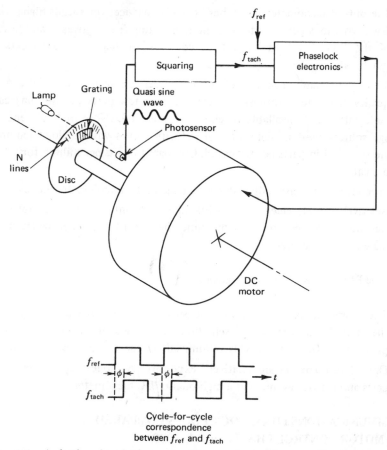

Figure 1-1 A basic phaselock servo. The motor/optics combination can be viewed as a voltage (or current) controlled oscillator. The resulting output frequency (f_{tach}) is compared to a reference frequency (f_{ref}), and the electronics causes a cycle-for-cycle correspondence between f_{ref} and f_{tach}.

The fundamental reason for the superiority of this system is that the optical tachometer, used as the velocity sensor, is capable of much better performance than the generator type of tachometer. When the optical disc is properly mounted on the motor shaft, it generates a frequency directly proportional to motor speed. Changes in air gap, temperature, and magnet strength simply have no effect on the output of the optical tachometer. By contrast, an analog tachometer is directly affected by all the problems listed above.

Motor speed control by phaselock has the following advantages:

1 The optical tachometer is the best speed transducer for speeds higher than a few revolutions per minute. It is noncontacting (i.e., no wear), and has none of the "speed voltage" problems common to magnetic transducers (i.e., synchros).

2 Very stable and accurate reference sources are readily available. A high quality frequency synthesizer (accurate to a few parts per million) can be built with readily available IC chips and a crystal. (Note that reference voltage sources used in conventional velocity servos with analog tachometers are specified in parts per thousand. Compare to parts per million for a crystal oscillator).

3 Speed is continuously adjustable by changing the reference frequency.

4 Relatively simple electronics yields all the advantages of a DC motor (high starting torque, control over dynamic characteristics), yet provides the ultimate in speed accuracy.

The PLL has two primary disadvantages:

1 These servo loops are somewhat sensitive to parameter changes. Changes in the inertial load may cause oscillation, requiring some readjustment of loop parameters. This difficulty is minimized by the techniques presented here.

2 Optical tachometers are relatively expensive. A modest optical tachometer costs about twice as much as a high quality DC tachometer.

COMMUNICATIONS PHASELOCK LOOPS COMPARED TO MOTOR CONTROL PHASELOCK LOOPS

The PLL has often been described and analyzed in the literature. These analyses most often pertain to communications loops, however. There are substantial differences between the PLL used for communications loops and those used for motor speed control. In the motor control situation, the motor/optics combination can be viewed as a VCO (voltage controlled oscillator) with inertia! VCOs with inertia have not been treated in the literature. Whereas the VCO in a conventional communications loop can change frequency very rapidly on command, the motor/optics VCO cannot. This factor greatly complicates the "lockup" process of the servo.

Consider the VCO used in communications loops. It has the transfer function

$$f_0 = kv \qquad\qquad (1\text{-}1)$$

where f_0 = output frequency (Hz)
   k = constant (Hz/V)
   v = input voltage (V)

No energy storage is present, and the VCO can theoretically change frequency instantaneously.

When a motor/optics combination is used as a current controlled oscillator, its transfer function is

$$\frac{f_{\text{tach}}}{I} = \frac{NK_T}{2\pi} \frac{1}{Js + K_D} \qquad (1\text{-}2)$$

where f_{tach} = output frequency of the tachometer (Hz)
   I = motor current (A)
   K_T = motor torque constant (oz-in/A)
   s = Laplace operator
   J = inertia (oz-in-sec^2)
   K_D = motor damping constant (oz-in./rad/sec)
   N = disc density

An extra pole is introduced because of energy storage in the system inertia. This extra pole complicates all aspects of the resulting loop. Furthermore, the pole location depends on the particular inertial load in use, which varies from one application to the next.

The generalized block diagram for the motor control PLL is shown in Fig. 1-2. Three error signals are generated: velocity error (at point A), integrated velocity error (at point B), and doubly integrated velocity error (at point C). The error signals are summed in Σ_4, amplified by transconductance amplifier A_1, converted to shaft rotation by the motor, and fed back through optical tachometer N. A_1 is a transconductance amplifier that converts input voltage to output current. M is a permanent magnet DC motor. The combination of the motor and optical tachometer may be considered a current controlled oscillator, because the motor is being driven by current source A_1.

It is immediately evident that this motor control PLL differs in many respects from the conventional PLL. This block diagram and corresponding servo system were evolved over a period of time. The reasons for this exact configuration are manyfold. One especially important consideration is the inability to produce an adequate phase detector for motor speed control applications. (See Chapter 3 on phase detectors.) Further insight into why the block diagram is as shown can be found in Appendix B and Chapter 5.

In communications loops the error detector is a true phase detector whose output is proportional to the phase difference between the reference and tachometer frequencies. Phase detectors are used to recover the information contained in the phase of the reference signal. Mathematically stated, a true phase detector provides

$$\epsilon = \theta_{ref} - \theta_{tach} \qquad (1\text{-}3)$$

where ϵ = phase error. In motor control loops there is no information in the phase of the reference signal. Consequently, motor speed controls do not require phaselock, but only integration of the frequency error. This is a somewhat less stringent condition on the error detector. A simple integrator provides

$$\epsilon = \int (\omega_{ref} - \omega_{tach})dt + K \qquad (\omega = 2\pi f) \qquad (1\text{-}4)$$

This integrator, when used in the PLL, forces the error between f_{ref} and f_{tach} to zero, thereby producing "frequency lock." However, the integrator develops no phase correspondence between f_{tach} and f_{ref}, because K is not specified, and may even drift! Phase correspondence is produced by auxiliary means. A number of different types of error detectors are available, each with its own merits and liabilities. These are fully described in Chapter 3. An overview is presented here.

1 *The Product Detector* The signals to be phaselocked are multiplied together (to get the product of the signals) and fed through a low pass filter. This yields quite acceptable results for communications loops, but is useless in motor speed control applications. The reason for its unacceptability is that it produces no DC for acceleration or deceleration when out of lock. Furthemore, even if the motor is accelerated by auxiliary means, lockup may occur on harmonics, thereby producing incorrect shaft speeds.

2 *The "Phase-Frequency" Detector* This detector does provide DC for acceleration and deceleration, it is immune to harmonic locking, but exhibits a severe nonlinearity in the frequency domain. This nonlinearity greatly complicates lockup, and the phase-frequency detector is therefore not preferred for motor control applications.

3 *The Counter and D/A "Detector"* This is not a true phase detector, but an integrator. It uses a digital up-down counter (often 8 bits long) and a D-to-A

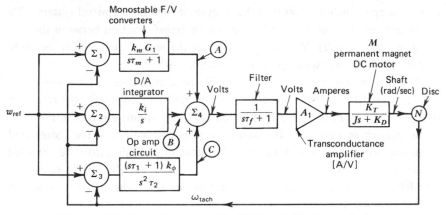

Figure 1-2 Block diagram of the PLL suitable for analysis.

converter (also 8 bits). It provides an output of the form of eq. 1-4 and helps the designer overcome the problems inherent in the other phase detectors.

CHOOSING THE INDEPENDENT VARIABLE: θ OR ω?

The servo system of Fig. 1-2 can be considered from two viewpoints: as a true phaselock servo with phase θ as the independent variable or as a "frequency locked" servo with ω as the independent variable.

1 The "phase" servo (θ is the independent variable) aligns the tachometer phase to the reference phase. When the phases differ by a constant, the frequencies are, by definition, identical, and a cycle-for-cycle correspondence between f_{tach} and f_{ref} exists. This viewpoint is widely used in the literature, but has substantial drawbacks for motor control work.

2 The "frequency" servo (ω is the independent variable) has a pure integration in the signal path. The pure integration reduces the error between f_{ref} and f_{tach} to zero. Again, a cycle-for-cycle correspondence between tachometer and reference frequencies exists, but the relative phase is random. The phase is then controlled by auxiliary means. This is the preferred approach for motor control.

When a mathematical analysis is made, both approaches yield seemingly identical results. Whether the describing equations are written as a function of θ

or of ω is apparently irrelevant to the analysis of the motor control system. This seems particularly true in view of the simple transformation between the two variables ($\theta = \int \omega \, dt$). Yet, it turns out that the second approach is the more fruitful one for motor speed control applications. Why?

The first reason is that there is generally no significance to the phase of the reference oscillator. There is no point in detecting or following the reference oscillator phase! Eliminating that unnecessary constraint eliminates the need for a true phase detector. The second reason for preferring the "integrated frequency error" approach is that the attributes of error detectors are obscured when θ is the variable and highlighted when ω is the variable. When ω is used, the defects of conventional phase detectors become evident. This is not so when θ is used as the variable.

To summarize this introduction:

1 Conventional velocity servos are limited in accuracy by the necessity for some speed error.
2 Speed error can be eliminated by using an integration in the forward path. However, the many limitations of the analog tachometer prevent optimum performance. Time, temperature, normal shifts in air gap, and magnet strength, as well as machanical wear, all change the transfer function of the tachometer and degrade initial accuracy.
3 With an optical tachometer and a motor control PLL, state-of-the-art speed control can be achieved.
4 Conventional PLL analyses are not directly applicable to the motor control PLL. This is because of the inertia always present in the motor control system and the differences in phase detectors required.
5 No suitable phase detector for the motor speed control PLL has been found; however, a digital integrator, with auxiliary phase control circuitry, can do the job of the phase detector.
6 The motor control PLL should be analyzed in terms of ω (radian frequency) rather than θ (phase angle).

A final note. The designations f for frequency (Hz) and ω for frequency (rad/sec) are often used to specify the same variable in the same paragraph. Using f(Hz) is advantageous when doing practical work. It is easy to read the frequency from a scope trace by simply inverting the period T of the signal. Using ω(rad/sec) is advantageous for analytic work. It avoids the need to carry a 2π factor around. (Recall that $\omega = 2\pi f$.) These interchanges are sometimes made in this book. All equations have the 2π factor, when appropriate.

2

Optical Tachometers

BASIC CONCEPTS

Optical tachometers are used as the feedback element for phaselocked systems. Their low inertia, low noise, high resolution, and high accuracy facilitate control of DC machines. The optical tachometer is the most critical component of the PLL system. Its quality determines the instantaneous speed accuracy at the shaft. The basic optical tachometer is shown in Fig. 2-1. A disc with N lines is mounted on the motor shaft. N is typically between 100 and 5000 lines (for a 2½ in. diameter disc). A grating that has the same line density as the disc is mounted over the sensor element, which is often a phototransistor or solar cell. As the disc rotates once, N lines are swept past the detector. The frequency of the output signal then is

$$f_{\text{tach}} = \frac{N\,\text{rpm}}{60} = \frac{N\omega}{2\pi} \tag{2-1}$$

where f_{tach} = tachometer output frequency (Hz)
N = disc density
rpm = revolutions per minute of motor shaft
ω = shaft angular velocity (rad/sec)

The grating is necessary whenever the sensor diameter is large compared to the line spacing of the disc. Such a case is depicted in Fig. 2-2a. As the disc rotates CW, the leftmost line starts to pass over the sensor, while the rightmost line leaves the area over the sensor. The net change in light intensity is very small, if not zero. This is because the average light intensity is almost unchanged as

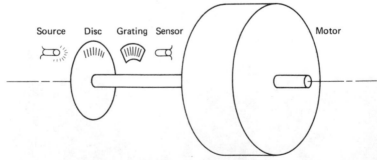

Figure 2-1 Exploded view of a basic optical tachometer. Shaft length between motor and disc is exaggerated for clarity.

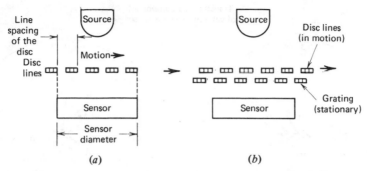

Figure 2-2 (*a*) As one line moves over the sensor from the left, another line moves off the sensor on the right. The net change in light intensity on the sensor is zero. (*b*) The grating forms individual slits across the face of the sensor.

one line replaces another. The addition of a grating creates individual slits across the face of the sensor. These allow almost complete obscuration or maximum illumination, depending on the relative position of disc and grating. (See Fig. 2-2*b*.) For low density discs ($N \leqslant 500$) the grating is often omitted because sensor diameter is small compared to disc line spacing.

A number of commercial vendors produce "interruptor" modules (Fig. 2-3). These consist of an LED/phototransistor combination prealigned and ready for use. The phototransistor is behind a single slit approximately 0.020 in. wide. A grating is not necessary when these interruptors are used with discs that have 200 to 300 lines.

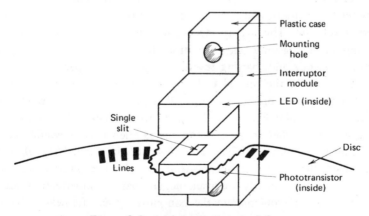

Figure 2-3 An interruptor module.

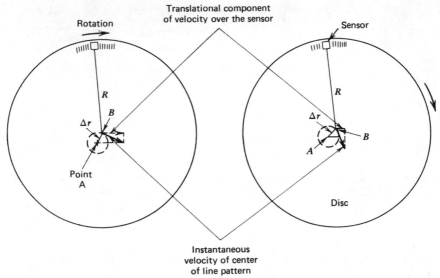

Figure 2-4 When the disc is misaligned, the rotation occurs about point *A*, but the center of the line pattern is at point *B*. As a consequence, the line pattern has both a translational and a rotational component of velocity across the sensor. R = radius of line pattern. Δr = Distance between center of line pattern and center of rotation.

ERROR SOURCES IN OPTICAL TACHOMETERS

It should first be emphasized that the accuracy of the system over an integral number of revolutions is essentially independent of the tachometer alignment. The disc closes on itself, and N cycles of output frequency are always generated in every revolution. The average speed accuracy over an integral number of revolutions (including only one revolution) is as good as the accuracy of the reference frequency and, barring severe problems, is independent of the disc.

Imperfections in the optical tachometer manifest themselves as variations during a single revolution. For example, if the lines on the disc were unevenly distributed on its periphery and the shaft were rotated, the denser regions would generate a higher frequency and the rarefied regions would generate a lower frequency, indicating higher and lower speeds, respectively. However, as just described, the average frequency over the whole revolution has no error.

Discs of excellent quality are available, and pattern imperfections are avoidable. Two common problems are the concentricity and flatness of the disc in relation to the surrounding structure. If the center of the disc line pattern is not mounted concentrically with the axis of rotation, frequency modulation of the output occurs producing an effect similar to line density variation. This can be seen in Fig. 2-4.

Instead of rotating about the center of the line pattern on the disc, the shaft center is displaced from the center of the line pattern by Δr. The linear velocity of the disc pattern over the sensor is increased (or decreased), and the zero crossings of the quasi sine wave are speeded up (or delayed). The expression for the output frequency is shown to be

$$f_{\text{tach}} = \frac{\omega N}{2\pi}\left(1 + 2\frac{\Delta r}{R}\sin\theta\right)$$ (2-2)

where R = pattern radius
f_{tach} = output frequency of tachometer
ω = angular velocity of shaft
θ = shaft angle = ωt
Δr = distance between the center of the disc pattern and the center of rotation.

Note that if $\Delta r = 0$ (disc is concentric) there is no FM. If $\Delta r \neq 0$, an FM component with the frequency of the shaft angular velocity modulates the nominal frequency. The derivation of eq. 2-2 is now presented. Please refer to Fig. 2-5 as needed.

CONCENTRICITY ERROR ANALYSIS

The time T for one complete cycle of output frequency is

$$T = \frac{\text{linear distance of a line-space pair}}{\text{linear speed of the track over the sensor}}$$ (2-3)

The line pattern is assumed to be perfect. It follows that the linear distance of a line-space pair is

$$\text{distance of a line-space pair} = \frac{2\pi R}{N}$$ (2-4)

where R = radius of the pattern
N = number of lines on the disc

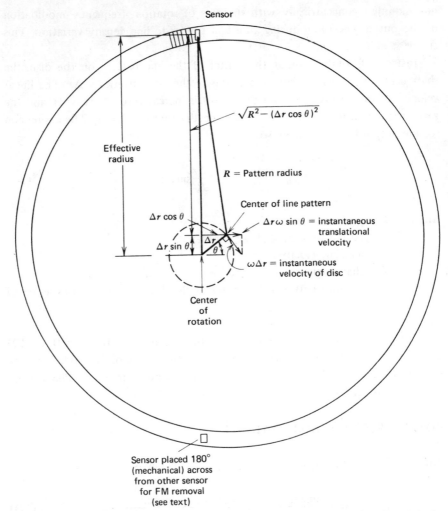

Figure 2-5 Diagram of the optical disc for concentricity error analysis. Δr = distance between center of pattern and center of rotation; ω = angular velocity.

This equation arises from the fact that there are N lines along a linear distance of $2\pi R$.

The linear speed of the pattern over the sensor can be resolved into two components. One is the velocity component due to the rotation of the shaft at ω rad/sec. The other is the component due to the translation of the pattern because of its misalignment. The translation occurs because the center of the pattern moves with respect to the center of the shaft. The rotational velocity component may be written as

$$\text{velocity (due to rotation)} = \omega \times \text{effective radius} \qquad (2\text{-}5)$$

From Fig. 2-5, the effective radius is

$$\text{effective radius} = \sqrt{R^2 - (\Delta r \cos \theta)^2} + \Delta r \sin \theta \qquad (2\text{-}6)$$

Substituting eq. 2-6 into eq. 2-5 and factoring R from the radical:

$$\text{velocity (due to rotation)} = \omega \left\{ R \sqrt{1 - \left(\frac{\Delta r}{R}\right)^2 \cos^2 \theta} + \Delta r \sin \theta \right\} \qquad (2\text{-}7)$$

This represents one component of the linear velocity of the pattern over the sensor. An additional component due to the translational movement of the center of the pattern is considered next.

The component of translational velocity that affects the sensor is

$$\text{translational component} = \Delta r \omega \sin \theta \qquad (2\text{-}8)$$

Equation 2-3 may now be written as

$$T = \frac{2\pi R/N}{\omega[R\sqrt{1 - (\Delta r/R)^2 \cos^2 \theta} + \Delta r \sin \theta] + \Delta r \omega \sin \theta} \qquad (2\text{-}9)$$

This is inverted (to obtain f) and rewritten as

$$f_{\text{tach}} = \frac{1}{T} = \frac{\omega N}{2\pi} \left[\sqrt{1 - (\Delta r/R)^2 \cos^2 \theta} + \frac{2\Delta r}{R} \sin \theta \right] \qquad (2\text{-}10)$$

In all cases of practical interest, $\Delta r/R \ll 1$. Typically $R = 1$ in., $\Delta r = 0.001$ in., and $(\Delta r/R)^2 = 10^{-6}$. Since the maximum value of $\cos^2 \theta$ is 1, the entire term inside the radical is approximately 1. Equation 2-10 becomes

$$f_{\text{tach}} = \frac{\omega N}{2\pi} \left(1 + \frac{2\Delta r}{R} \sin \theta\right) \qquad \text{for } \frac{\Delta r}{R} \ll 1 \qquad (2\text{-}11)$$

and

$$\omega_{tach} = \omega N \left(1 + \frac{2\Delta r}{R} \sin \theta \right) \qquad (2\text{-}12)$$

where ω_{tach} = radian frequency of the tachometer
ω = radian speed of the shaft

ERROR CANCELLATION

The FM component of the waveform can obviously be reduced by improved alignment of the disc (i.e., decreasing Δr). Another means of further reducing the FM is to introduce an additional sensor displaced 180° from the first sensor (i.e., directly opposite the first sensor). This is shown in Fig. 2-5. The outputs of the two sensors are summed, and the resulting signal has substantially less FM than the originals.

The rationale behind this scheme is that when the speed of the disc is increased (because of misalignment) over one of the sensors, it is decreased over the other sensor. These signals, when summed, tend to cancel the effect of the misalignment. A more analytic approach uses the results of the previous derivation.

It is first necessary to have an expression for the output voltage of the original sensor. Note that the previous derivation resulted in an expression for the frequency of the sensor signal, and not its output voltage. It is assumed that the output voltage is a sinusoid. In order to obtain an expression for the output voltage, the frequency must be converted to an angle, because the argument of a sinusoidal function must be an angle. To accomplish this conversion, use is made of two relationships:

$$\text{phase angle} = \int \omega_{tach} \, dt \qquad (2\text{-}13)$$

and

$$\theta = \omega t \qquad (2\text{-}14)$$

where θ = shaft position (rad)
ω = shaft speed (rad/sec)
t = time (sec)

Equation 2-12 is repeated below for reference as

$$\omega_{\text{tach}} = \omega N\left(1 + \frac{2\Delta r}{R}\sin\omega t\right)$$ (2-12)

Substituting eq. 2-12 into eq. 2-13 yields

$$\text{phase angle} = \int \omega_{\text{tach}}\,dt = \int \omega N\left(1 + \frac{2\Delta r}{R}\sin\omega t\right)dt$$ (2-15)

Evaluating the integral:

$$\text{phase angle} = \omega Nt + \alpha$$

where

$$\alpha = -\frac{2\Delta r}{R}N\cos\omega t.$$ (2-15a)

The output voltage of the sensor can then be written as

$$v_o(t) = V_{\text{max}}\cos(\omega Nt + \alpha)$$ (2-16)

This is the sinusoidal output voltage of the original sensor. The FM is introduced by α. Note that α is time dependent.

If an additional sensor is placed 180° mechanical (across) from the sensor described above (see Fig. 2-5), it can be shown that this 180° sensor will develop an output voltage ($= v_{180}(t)$) of

$$v_{180}(t) = V_{\text{max}}\cos(\omega Nt - \alpha)$$ (2-17)

Note the change in sign of phase α. If the two signals are added, the phasor diagram of Fig. 2-6 applies. The frequency modulation has been removed. This technique is often used to improve the performance of the optical tachometer. Ideally, the FM component disappears. In practice some FM remains because of the harmonic content of the optics signals.

Consider next the problems associated with the disc not mounted in a plane exactly perpendicular to the shaft centerline. Wobble results, as illustrated in Fig. 2-7. The wobble introduces amplitude modulation of the output because the illumination of the sensor depends not only on disc rotation, but disc position. Consider Fig. 2-8 where a side view of the disc, grating, and sensor

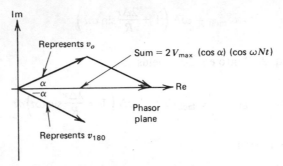

Figure 2-6 Removal of FM can be accomplished by summing the outputs of two sensors located 180° apart on the disc. The sum v_o and v_{180} produce a signal with no FM, as shown on the phasor diagram.

are shown. If the disc wobbles, the light on the sensor is modulated by the "variable distance" motion of the disc producing a "once around" component. It is observed as an amplitude modulation on the output sine wave. Two steps can be taken to overcome this problem. First, the hub on which the disc is mounted is "secondary machined," that is, machined square to the shaft after it has been mounted on the motor. Second, another optics pickoff (the same one used for FM cancellation) also helps with disc wobble. This is because opposite effects occur at either side of the disc and the two signals from the opposite "sides" tend to cancel each other.

It is seen how mounting of the disc is a primary determinant of output signal quality. Furthermore, it is seen how the addition of a second pickoff helps to eliminate mounting errors.

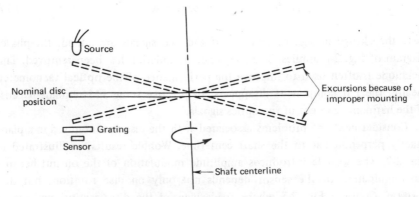

Figure 2-7 Side view of a disc with wobble.

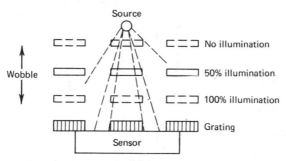

Figure 2-8 The intensity of light on the photosensor is modulated by wobble.

IMPROVING PHOTOTRANSISTOR BANDWIDTH

An inexpensive and very effective light sensor is the silicon phototransistor. Its spectral response is high at the emission wavelengths of both LEDs and incandescent lamps. The output signal level of the phototransistor is on the order of a few hundred millivolts to as low as 1 mV under certain operating conditions. Furthermore, a relatively wide bandwidth is often required from the sensor circuit to accommodate the combined effect of a high density disc turning at high speed. For example, a 5000 line disc running at 3000 rpm produces an output frequency of

$$f = \frac{N \text{ rpm}}{60} = \frac{5000 \times 3000}{60} = 250 \text{ kHz}$$

To achieve this speed of response, it is necessary to use circuits that "look" like a short circuit to the phototransistor, thereby effectively "shorting out" the internal capacitances of the sensor. The phototransistor and its associated circuitry are now analyzed.

A common way of using a phototransistor is shown in Fig. 2-9. The light input is equivalent to base current, and the output consists of a DC value plus the signal. It is desired to obtain maximum bandwidth in the response of this phototransistor circuit. The nature of the phototransistor is such that the best frequency response is obtained at very low values of load resistance. In fact, zero load resistance yields the best frequency response. However, as the load resistor decreases, the magnitude of the output voltage also decreases. This leads to a conflict, requiring a high value of load resistor for signal amplitude and a low value for best bandwidth.

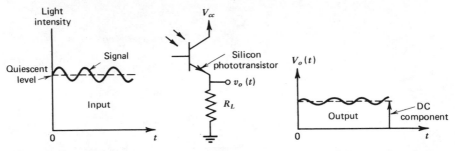

Figure 2-9 The light intensity signal shown at the left falls on the photo-transistor. The resulting output voltage is shown at the right.

The reason why a low load resistor is needed for wide bandwidth can be appreciated by examining the equivalent circuit of the phototransistor in Fig. 2-10. For the voltage to rise rapidly at the emitter, the current generator $g_m V_{be}$ must charge C_{ce}. This charging action has associated a time constant, RC_{ce}, where R is the parallel combination of R_L and R_{ce}. As R_L decreases, so does the time constant RC_{ce}, until, when $R = 0$, $V_o = 0$ and no charge at all must be delivered into C_{ce}. Thus the fastest response is with $R_L = 0$.

A more exact analysis of this circuit can be carried out by using the equivalent circuit, which shows that

$$V_o \approx \frac{R\beta I_\lambda}{(1 + j\omega RC_{ce})(1 + j\omega r_{be}C_{be})} \tag{2-18}$$

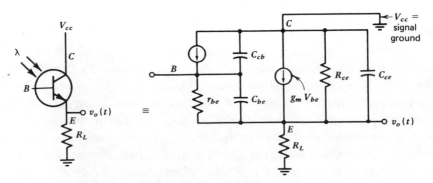

Figure 2-10 Equivalent circuit for a phototransistor.

where R = parallel combination of R_L and R_{ce}
    I_λ = base current produced by the radiation
    V_o = output voltage
    R_L = load resistor
    β = AC current gain of the transistor

From eq. 2-18 output current (into R_L) is found to be

$$I_o = \frac{V_o}{R_L} \approx \frac{\beta I_\lambda}{(1 + j\omega R C_{ce})(1 + j\omega r_{be} C_{be})} \qquad (R_L \ll R_{ce}) \qquad (2\text{-}19)$$

By using a zero ohm load resistance ($R_L = 0$), eq. 2-19 becomes

$$I_o = \frac{\beta I_\lambda}{1 + j\omega r_{be} C_{be}} \qquad (2\text{-}20)$$

The time constant RC_{ce} has been eliminated. Since $RC_{ce} \gg r_{be} C_{be}$, the bandwidth is considerably improved. The key point here is that the *current* generated by the phototransistor is sensed rather than the voltage across an added resistor.

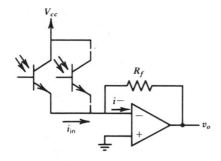

Figure 2-11   A circuit where the op amp presents a load impedance of 0 Ω (ideally) to the phototransistor.

One way of current sensing is to use the op amp as shown in Fig. 2-11. The net signal current into the inverting input of the op amp is nominally zero. As the phototransistor delivers current to the input of the op amp, the current is "drawn out" by v_o through the feedback resistor. This maintains the op amp input current at zero. For an ideal op amp, $i^- = 0$ and it follows that

$$i_{in} + \frac{v_o}{R_f} = 0 \qquad (2\text{-}21)$$

and

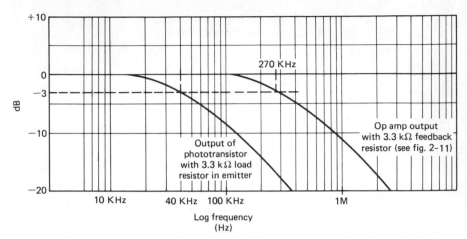

Figure 2-12   Output versus frequency for phototransistor circuit similar to that of Fig. 2-9 versus circuit of Fig. 2-11; 0 dB is referred to 1 kHz output of former circuit.

$$v_o = -i_{\text{in}} R_f \qquad (2\text{-}22)$$

The sensor is operated into what is effectively a short circuit (the virtual ground of the op amp), thereby "shorting out" the internal capacitances that limit sensor bandwidth. Any number of phototransistors can be paralleled at the op amp input. This is useful for FM cancellation schemes in optical encoders. A graph of the response of this circuit compared to using only a transistor with a load resistor is given in Fig. 2-12. The advantage gained is observed from the graph. The output stays completely flat to 100 kHz as compared to only 15 kHz for the simple transistor circuit.

A more detailed analysis of this op amp circuit can be made by writing the basic equations that govern its operation. (See Fig. 2-13a.)

$$\frac{v_o - \epsilon}{R_f} + i_{\text{in}} - \frac{\epsilon}{R_{\text{in}}} = 0 \qquad (2\text{-}23)$$

$$\epsilon A = v_o \qquad (2\text{-}24)$$

$$Z_{\text{in}} = \frac{\epsilon}{i_{\text{in}}} \qquad (2\text{-}25)$$

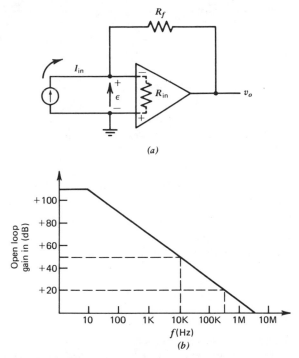

Figure 2-13 (a) Circuit for analysis of op amp performance in the nonideal case. (b) Typical op amp open loop gain as a function of frequency.

where i_{in} = phototransistor output current

A = open loop gain of the op amp

R_{in} = op amp input resistance

ϵ = input voltage to the op amp

R_f = feedback resistor

Using eqs. 2-23 and 2-24 and solving for v_o,

$$v_o = \frac{-R_f i_{in}}{1 - 1/A - (R_f/R_{in})(1/A)}$$

(2-26)

For an ideal op amp, $A \to \infty$ and

$$v_o = -R_f i_{in}$$

(2-22)

Using eqs. 2-24 and 2-25, it follows that

$$Z_{in} = \frac{v_o}{A i_{in}} \tag{2-27}$$

Substituting into eq. 2-26, it follows that

$$Z_{in} = \frac{-R_f}{A - 1 - R_f/R_{in}} \approx \frac{-R_f}{A} \quad (A \text{ is a negative number}) \tag{2-28}$$

Equation 2-28 allows the designer to predict the "quality" of the virtual ground. For example, consider the open loop gain of a type LF347 op amp whose open loop transfer function is shown in Fig. 2-13b. At 10 kHz, the open loop gain is approximately 50 dB, so that $A \cong 316$. Assuming a value of 3.3K for R_f and using the approximation for Z_{in}, we have

$$Z_{in} \approx \frac{R_f}{A} = \frac{3.3K}{316} \approx 10\,\Omega \tag{2-28}$$

The output voltage is being developed across a 3.3K resistor, yet the transistor "sees" only $10\,\Omega$ resistance! At 250 kHz, $A \approx 20$ dB or a gain of 10. At that frequency

$$Z_{in} \approx \frac{R_f}{A} = \frac{3.3K}{10} = 330\,\Omega$$

Clearly, the performance is degraded, but the transistor develops an output across 3.3K while "seeing" only $330\,\Omega$.

A variation of this circuit, shown in Fig. 2-14 includes an integrator that automatically removes the DC component of the output signal. If f_{min} is defined as the lowest input frequency to be used, RC is chosen as

$$RC \geqslant \frac{1}{10\pi f_{min}} \tag{2-29}$$

The circuit is absolutely stable and yields the same performance as the previous circuit for all frequencies greater than f_{min}. The DC component of the output is nominally zero.

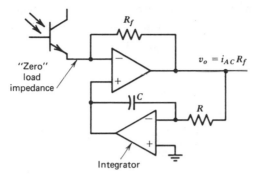

Figure 2-14 A variation of the circuit of Fig. 2-11. The DC component in the output is automatically removed.

LIGHT SOURCES

Tungsten lamps have many advantages over LEDs. The claim of long life is a much praised attribute of LEDs. However, incandescent lamp life varies inversely as the 12th power of the ratio of the applied voltage versus the rated voltage. Typically, 50,000 hours of continuous operation (i.e., about 5 ½ years) can be obtained by running incandescent lamps 10% derated. This is accomplished by including a resistor in series with the lamp. When the nominal voltage is applied, the lamp runs properly derated. Furthermore, lamp light is visible, which greatly aids in alignment procedures, and lamp output is not temperature sensitive, as is LED output. Lamp (or LED when used) supply voltage must be regulated to ensure that there is no unwanted modulation due to current variations in the lamp.

Some additional factors should be considered when building or ordering motors with optics. The bearing structure that holds the shaft in place must be such that the expected loads on the machine do not push, pull, or bend the optics disc out of "true." Class 7 bearings, with provision to prevent shaft end play, are normally required.

To obtain optimum performance from DC motor optics, take the following steps as needed:

1   Properly mount disc to shaft, thus minimizing runout and wobble.
2   Include a second sensor located 180° mechanical degrees from the first sensor, with the two signals added. This compensates for remaining mounting errors.

3 Use derated tungsten lamps for ease of alignment and long life.
4 Process signal as close to the motor as possible to avoid noise contamination.
5 Use proper bearings on the motor shaft.

3

Information Recovery from the Optical Tachometer

This chapter outlines some techniques for obtaining velocity and direction information from the optical tachometer. In addition, methods for detecting the phase error between f_{ref} and f_{tach} are presented. It is shown that none of the commonly available phase detectors is suitable for the motor control PLL. Instead, a digital integrator is used to reduce the frequency error to zero. The phase is adjusted by auxiliary means as described in Chapter 5.

VELOCITY SENSING

A simple velocity sensing circuit can be realized with a monostable multivibrator. The "mono" or "one shot" is connected as a "frequency-to-voltage" converter as shown in Fig. 3-1a. To achieve frequency-to-voltage conversion, the squared optics signal is fed to the TR^+ input of the mono. Every positive-going input edge on input TR^+ causes a uniform width pulse (width $= \tau$) at the Q output of the mono.

When the motor rotates at a low speed, the tachometer frequency is relatively low and the positive edges to TR^+ appear relatively infrequently. This is illustrated in Fig. 3-1b. At higher speeds, positive edges occur more frequently, as shown in Figs. 3-1c and d. The RC network at the output of the monostable extracts the DC component of the output waveform. The DC component of the output increases with speed because of the greater density of output pulses at high speed. Assuming that the one shot output has a high level output of V_{cc} and a low level output of 0 V (typical of CMOS devices), it can be written that

$$V = \frac{\tau}{T} V_{cc} \tag{3-1}$$

where τ = period of monostable (sec)
$\quad\quad T$ = period of tachometer signal (sec)
$\quad\quad V_{cc}$ = supply voltage (V)
$\quad\quad V$ = DC output component of monostable waveform

Furthermore, τ is often chosen so that at the maximum input frequency (i.e., maximum speed) the duty cycle of the mono output is 100%. Stated another way, when the maximum input frequency is applied, the period of the input signal equals the width of the monostable output. This choice yields maximum sensitivity over the required speed range. Stated mathematically:

28

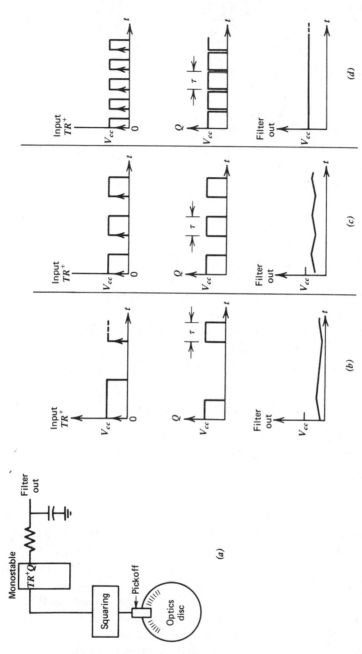

Figure 3-1 (*a*) A simple F/V (frequency-to-voltage) converter that uses an edge triggered monostable and an RC filter. (*b*) At low tachometer speed relatively few positive edges occur in a given time. The resulting DC component is small compared to V_{cc}. (*c*) At medium speeds a moderate number of positive edges produce a higher density of output pulses from the monostable. The DC component increases over the low speed value. (*d*) At speeds approaching the maximum range of the F/V converter the monostable output pulses are closely crowded together. The DC component is close to V_{cc}.

$$\tau = \frac{1}{f_{\text{max}}} \tag{3-2}$$

Using the relationship $T = 1/f$, it follows that

$$V = \frac{1/f_{\text{max}}}{1/f} V_{cc} = \frac{f}{f_{\text{max}}} V_{cc} \tag{3-3}$$

Converting to radian measure ($\omega = 2\pi f$):

$$\frac{V}{\omega} = \frac{V_{cc}}{\omega_{\text{max}}} \triangleq k_m$$

where k_m = gain of the F/V converter (V/rad/sec). A simple RC network is often used to filter the AC components of the monostable output. The time constant of the monostable output filter, RC, is selected as part of the servo design. Its selection depends on the lowest frequency that will occur, which in turn depends on disc line density N and the minimum speed requirement. Details of how to select RC are presented in Chapter 4.

The transfer function for the entire speed sensing system is

$$\frac{V}{\omega} = \frac{k_m}{s\tau_m + 1} \tag{3-5}$$

where $\tau_m = RC$ = filter time constant
$k_m = V_{cc}/\omega_{\text{max}}$

This basic circuit can be enhanced by having the monostable trigger from both the positive-going and negative-going edges of the optics waveform. This doubles the output frequency and thus halves the required time constant. This 2-to-1 improvement allows the use of a higher (i.e., faster) time constant in the filter, thus introducing less lag into the resulting servo system.

A serious disadvantage of this simple monostable speed sensor is that it is insensitive to direction. It produces the same output whether the motor runs CW or CCW, which makes it not generally useful for servo work. This limitation can be circumvented, however, as discussed in the sections that follow.

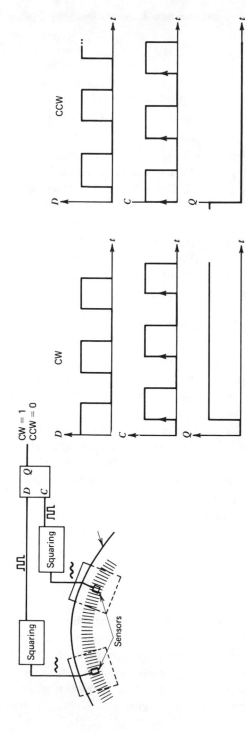

Figure 3-2 Diagram of a simple direction sensing system on the optical tachometer, with waveforms for CW and CCW rotation.

DIRECTION SENSING

A simple direction sensing system is implemented by using an additional optical sensor, a squaring circuit, and a D flip-flop. The additional optics sensor is placed mechanically adjacent to the original sensor and its output is adjusted to be 90° out of phase with the output of the original sensor. The system and its output waveforms are shown in Fig. 3-2.

In this configuration, CW rotation causes C to lag D by 90° and CCW rotation causes D to lag C by 90°. The reason is that edges that "rise" for CW rotation "fall" for CCW rotation and vice versa. The D flip flop transfers the input on the D terminal to the output Q terminal on each successive rising edge of the clock input signal. From Fig. 3-2 it is observed that signal C clocks the flip-flop and signal D constitutes the D input. For CW rotation, a logic 1 is always transferred to Q, for CCW rotation a logic 0 is transferred to Q. This simple direction sensing scheme is quite adequate for PLL servo applications.

FOUR QUADRANT VELOCITY SENSING

The summing junction in a servo system must be capable of four quadrant operation. That is, if either the reference signal or the feedback signal changes magnitude or polarity, the change must be transmitted as an error voltage of appropriate magnitude and polarity.

The simple monostable velocity sensing circuit described at the beginning of this chapter makes no polarity change when the direction reverses. It provides the identical DC voltage for both CW and CCW rotation. This is not adequate for servo work. A circuit that combines the monostable velocity sensor and the direction sensing signal from the D flip-flop is presented. It makes possible completely general servo operation that uses only the outputs from the optical tachometer and the processing previously described. The complete circuit is shown in Fig. 3-3a.

CIRCUIT DESCRIPTION

Note first that the logic signals for the "command direction" and the "actual direction" are "reversed." If it is desired that the motor turn CW, the direction command signal is set to 0. If and when the motor actually obeys this command signal and actually runs CW, the "actual direction" signal is 1. Similarly, if CCW rotation is desired, the direction command signal is set to 1. When CCW rotation exists, the "actual direction" signal is 0. During "normal" operation

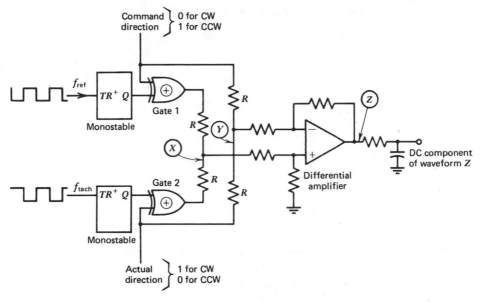

Figure 3-3a A four quadrant velocity error detector.

the "actual direction" signal is opposite the "commanded direction" signal. The four possibilities are shown in Chart 3-1. Note also that logic 1 is represented by voltage V_{cc}. Logic 0 is represented by 0 V. These levels are characteristic of the CMOS logic family. When the "actual direction" and "commanded direction" voltages are combined by superposition across two resistors R (Point Y in Fig. 3-3a), four possibilities exist, as shown in Chart 3-2. This chart is used later in the analysis.

An exclusive OR gate may be considered to be an "electronic inverter." If one input of an exclusive OR gate is maintained at logic 1, then the signal

CHART 3-1   DIRECTION SIGNAL LOGIC STATES

Command Direction	Actual Direction
0 (CW)	1 CW ("normal")
0 (CW)	0 CCW ("changing")
1 (CCW)	1 CW ("changing")
1 (CCW)	0 CW ("normal")

CHART 3-2   POINT Y IN FIG. 3-3a

Direction Command	Actual Direction	Voltage Output
0 (CW)	V_{cc} (CW)	$V_{cc}/2$
0 (CW)	0 (CCW)	0
V_{cc} (CCW)	V_{cc} (CW)	V_{cc}
V_{cc} (CCW)	0 (CCW)	$V_{cc}/2$

at the second input terminal appears inverted at the output of the gate. When logic 0 is applied to one terminal of the exclusive OR, the input to the other terminal is transmitted directly. Stated another way, a logic 1 applied to one input of the exclusive OR causes the gate to be an inverter to the other input signal. A logic 0 applied to one input of the exclusive OR causes the gate to be simply a transmitter of the other input signal.

Assume now that the "command direction" signal is zero (i.e., CW). The output of gate 1 is simply a pulse train whose DC component is proportional to speed. This follows the earlier analysis exactly, and we have the situation shown in Fig. 3-3b.

$$V_{ref} = V_{cc}\frac{\tau}{T_{ref}} = V_{cc}\tau f_{ref}$$

Direction
command = 0 (CW)

where τ = monostable period

$$f_{ref} = \frac{1}{T_{ref}}$$

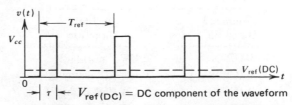

Figure 3-3b Output of gate 1 (Fig. 3-3a) for CW command. τ = monostable period; $f_{ref} = 1/T_{ref}$; V_{ref} (DC) = DC component of the waveform.

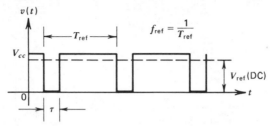

Figure 3-3c Output of gate 1 (Fig. 3-3*a*) for CCW command.

When the "command direction" signal is changed to logic 1, the exclusive OR gate becomes an inverter. The output waveform is shown in Fig. 3.3*c*, and the equation for the resulting output voltage is given as

$$V_{\text{ref (DC)}} = V_{cc} - V_{cc}\frac{\tau}{T_{\text{ref}}} = V_{cc}(1 - \tau f_{\text{ref}})$$
direction
command $= 1$ (CCW)

The signal is inverted, and the equation has a negative sign in front of f_{ref}.

Similarly, the output of gate 2, which is driven by the tachometer frequency and the "actual direction" signal, can be written as

$$V_{\text{tach(DC)}} = V_{cc}\tau f_{\text{tach}}$$
actual
direction $= 0$(CCW)

and

$$V_{\text{tach (DC)}} = V_{cc}(1 - \tau f_{\text{tach}})$$
actual
direction $= 1$(CW)

The two voltages (from the output of gates 1 and 2) are combined (across two resistors R) by superposition to form output X. The output voltage can be any one of four possibilities (Chart 3-3). The differential amplifier develops an output signal proportional to quantity $X - Y$. This quantity can be obtained by combining Charts 3-2 and 3-3 into Chart 3-4.

The results listed in Chart 3-4 show that the signal $X - Y$ represents a full four quadrant error signal between reference "velocity" (via f_{ref}) and actual velocity (via f_{tach}). In cases 1 and 4 the circuit acts as a frequency error detector. Since frequency and velocity are directly proportional, the output of

CHART 3-3 OUTPUT VOLTAGE AT POINT X FOR ALL LOGIC STATES

Case	Commanded Direction	Actual Direction	Output X $\left(= \dfrac{\text{Commanded} + \text{Actual}}{2}\right)$
1	$V_{\text{ref}} = V_{cc}\,\tau f_{\text{ref}}\,(\text{CW})$	$V_{\text{tach}} = V_{cc}\,(1 - \tau f_{\text{tach}})\,(\text{CW})$	$\dfrac{V_{cc}}{2}(\tau f_{\text{ref}} - \tau f_{\text{tach}} + 1)$
2	$V_{\text{ref}} = V_{cc}\,\tau f_{\text{ref}}\,(\text{CW})$	$V_{\text{tach}} = V_{cc}\,\tau f_{\text{tach}}\,(\text{CCW})$	$\dfrac{V_{cc}}{2}(\tau f_{\text{ref}} + \tau f_{\text{tach}})$
3	$V_{\text{ref}} = V_{cc}\,(1 - \tau f_{\text{ref}})\,(\text{CCW})$	$V_{\text{tach}} = V_{cc}\,(1 - \tau f_{\text{tach}})\,(\text{CW})$	$\dfrac{-V_{cc}}{2}(\tau f_{\text{ref}} + \tau f_{\text{tach}} - 2)$
4	$V_{\text{ref}} = V_{cc}\,(1 - \tau f_{\text{ref}})\,(\text{CCW})$	$V_{\text{tach}} = V_{cc}\,\tau f_{\text{tach}}\,(\text{CCW})$	$\dfrac{-V_{cc}}{2}(\tau f_{\text{ref}} - \tau f_{\text{tach}} - 1)$

CHART 3-4[a] VOLTAGE OUTPUTS OF THE FOUR QUADRANT VELOCITY SENSING CIRCUIT

Case	Command Direction	Actual Direction	X (from Chart 3-3)	Y (from Chart 3-2)	X − Y
1	CW	CW	$\frac{V_{cc}}{2}(\tau f_{ref} - \tau f_{tach} + 1)$	$\frac{V_{cc}}{2}$	$\frac{V_{cc}}{2}\tau(f_{ref} - f_{tach})$
2	CW	CCW	$\frac{V_{cc}}{2}(\tau f_{ref} - \tau f_{tach})$	0	$\frac{V_{cc}}{2}\tau(f_{ref} + f_{tach})$
3	CCW	CW	$\frac{-V_{cc}}{2}(\tau f_{ref} + \tau f_{tach} - 2)$	V_{cc}	$\frac{-V_{cc}}{2}\tau(f_{ref} + f_{tach})$
4	CCW	CCW	$\frac{-V_{cc}}{2}(\tau f_{ref} - \tau f_{tach} - 1)$	$\frac{V_{cc}}{2}$	$\frac{-V_{cc}}{2}\tau(f_{ref} - f_{tach})$

[a]The monostables are assumed to have identical periods.

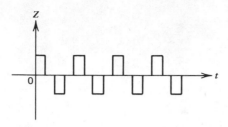
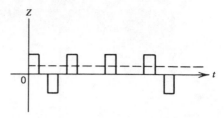

Figure 3-3d Output at point Z (Fig. 3-3a) when motor is phaselocked. Note that the net DC component is zero.

Figure 3-3e Output at point Z (Fig. 3-3a) when motor runs too slowly. Note that there is a net DC component to accelerate the motor.

the circuit represents the velocity error between the reference "velocity" and shaft velocity.

Case 2 occurs whenever the motor is running CCW and the direction switch is suddenly reversed, thereby commanding CW rotation. Both signals (from f_{ref} and f_{tach}) develop a positive voltage to accelerate the motor CW at the maximum rate. At some point, the directions again match and case 1 applies.

Similarly, case 3 occurs whenever the motor is running CW and the direction switch is suddenly reversed, thereby commanding CCW rotation. Both signals (f_{ref} and f_{tach}) combine to develop a negative voltage to accelerate the motor CCW at the maximum rate. When the directions match, case 4 applies. The output of the differential amplifier is followed by a simple RC filter, which passes the DC component of the $X - Y$ waveform and shunts the AC portion to ground.

Idealized output waveforms of the differential amplifier are shown in Figs. 3-2d and 3-2e. Refer first to Fig. 3-3d, which shows the output Z when f_{ref} and f_{tach} are identical. This occurs when the phaselocked condition is reached (by definition, $f_{ref} = f_{tach}$). For every positive-going pulse there exists a negative-going pulse, and the DC component of the output is zero. When the motor is running below "synchronous" speed, the waveform of Fig. 3-3e applies. On the average, there are more reference pulses than tachometer pulses. Consequently, a net DC component is available to accelerate the motor.

It is tacitly assumed that the periods of both monostables ($= \tau$) are identical. Although the periods can be adjusted to be equal, they depend on the stability of analog components, usually a resistor and a capacitor. Over the long term, the monostable periods must be expected to drift in varying amounts, thereby making the periods unequal.

Having different periods for each of the monostables modifies the error output in the same way as if one of the frequencies were slightly different

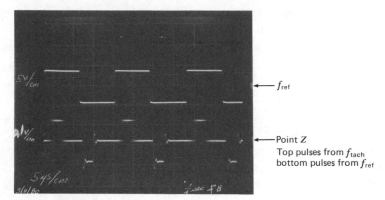

Figure 3-3f Photograph of f_{ref} and the signal at point Z (Fig. 3-3*a*) under locked conditions and CW rotation.

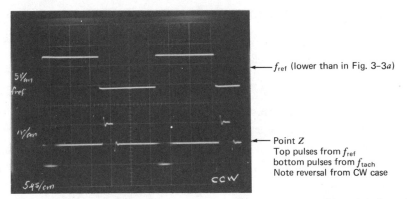

Figure 3-3g Photograph of f_{ref} and the signal at point Z (Fig. 3-3*a*) under locked conditions and CCW rotation.

from what it actually is. For example, if the period of the tachometer monostable increases, the DC component (for a given tachometer frequency) is greater than if τ had remained at its set time. A proportional velocity error appears at the motor shaft. This is true of any analog system. An analog tachometer can drift because of magnetic field strength changes, changes in the air gap of the device (i.e., the flux path), and other factors.

When the circuit described above is used in a PLL, the drifts are of no consequence! This is because the integrator (always present in a motor control PLL) automatically makes up for any steady state errors of the velocity detector circuit. By integrating the error between f_{ref} and f_{tach}, a synchronous condition is reached independent of the velocity error detector.

This is not to say, however, that the velocity error detector is less than an essential part of the PLL; it is essential! It is the dynamic performance of the velocity error detector that is necessary for proper loop operation, and this circuit provides the required four quadrant dynamic performance.

Figures 3-3f and g show output Z in relation to the reference frequency and under phaselocked conditions. (It is difficult to photograph Z when there is no phaselock, because the tachometer and reference pulses are not synchronized.)

The ringing on the pulses is due to the bandwidth limitation of the differential amplifier. The blurriness of the tachometer pulses is due to phase jitter of the tachometer output. The blurriness obscures the ringing on the tachometer pulses. Figure 3-3f shows CW rotation (reference pulses positive). Figure 3-3g shows CCW rotation (reference pulses negative).

PHASE DETECTORS–INTRODUCTION

To achieve zero speed error at the motor shaft, it is necessary to implement one of the following design schemes:

1   Use of a "traditional" PLL that detects differences in the phases of ω_{ref} and ω_{tach}. The resulting error signal has the form $\theta_{ref} - \theta_{tach}$. This follows the well known techniques for PLL analysis.

2   Use of a frequency locked loop scheme that integrates the error signal $\omega_{ref} - \omega_{tach}$. Auxiliary means are used to adjust the phase angles to some arbitrary relation.

The problem with approach 1 is that none of the commonly available phase detectors can work consistently in a motor control system. Approach 2 yields far better results. To understand this, it is helpful to review some of the well known phase detection methods.

PHASE DETECTORS–MULTIPLIERS

Phase detection can be accomplished by means of a multiplier. The multiplication may be a pure analog process, which actually multiplies two signals, or a switching process, which generates product terms. Both types are analyzed here.

Consider first the pure analog multiplication of two sinusoidal waveforms as depicted in Fig. 3-4. Such a process forms the basis of all the multiplier-type

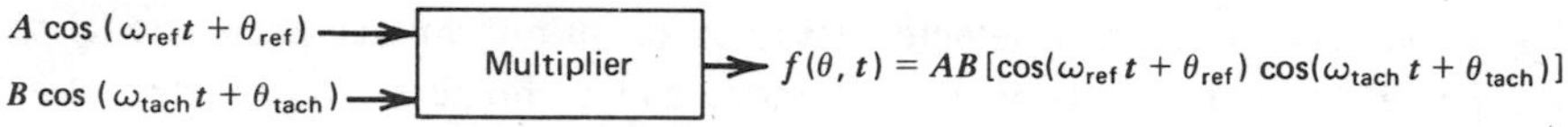

$A \cos{(\omega_{ref}t + \theta_{ref})} \longrightarrow$

Multiplier $\longrightarrow f(\theta, t) = AB\,[\cos(\omega_{ref}t + \theta_{ref})\cos(\omega_{tach}t + \theta_{tach})]$

$B \cos{(\omega_{tach}t + \theta_{tach})} \longrightarrow$

Figure 3-4 An analog multiplier used as a phase detector.

phase detectors, except that in other types additional harmonics are generated by the process or exist in the input signals. Using the trigonometric identity

$$\cos x \cos y = \frac{\cos(x+y) + \cos(x-y)}{2} \qquad (3\text{-}6)$$

we write

$$f(\theta, t) = \frac{AB}{2}\,[\cos((\omega_{ref} + \omega_{tach})t + \theta_{ref} + \theta_{tach})$$

$$+ \cos((\omega_{ref} - \omega_{tach})t + \theta_{ref} - \theta_{tach})] \qquad (3\text{-}7)$$

If the loop is assumed to be near lock, ω_{ref} and ω_{tach} are almost equal. Accordingly, it can be approximated that $(\omega_{ref} + \omega_{tach}) = 2\omega_{ref}$. One of the tasks of the carrier filter would be to remove this "double frequency" term. This presents no great difficulty, since $2\omega_{ref}$ is generally far above the loop bandwidth.

It remains to consider the difference term $(\omega_{ref} - \omega_{tach})$. Assuming (for the moment) a locked condition, $\omega_{ref} = \omega_{tach}$, and the output of the multiplier is written as

$$f(\theta, t) = f(\theta) = \frac{AB}{2}\cos(\theta_{ref} - \theta_{tach}) \qquad (3\text{-}8)$$

A sketch of $f(\theta)$ in Fig. 3-5 shows the relationship between phase error and

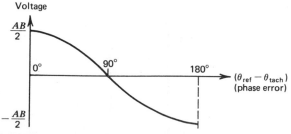

Figure 3-5 Multiplier output voltage as a function of phase difference between two sinusoidal input signals. The servo is assumed to be locked.

output voltage. This error detector develops an output voltage proportional to the phase difference between the reference and tachometer frequencies. Near $90°$, it is almost a linear relationship.

The analysis above assumes that the system is at, or at least near, lockup. Assume next that the system is not in phaselock and that $\omega_{ref} \neq \omega_{tach}$. The difference term $\cos[(\omega_{ref} - \omega_{tach})t + \theta_{ref} - \theta_{tach}]$ now represents the beat frequency between the tachometer and reference frequencies. Under these circumstances the error voltage, $f(\theta,t)$ delivered to the PLL consists of AC terms only. One term is the low frequency signal $\cos(\omega_{ref} - \omega_{tach})t + \Delta\theta$ and the other is the high frequency signal $\cos(\omega_{ref} + \omega_{tach})t$. Neither supplies any net DC to drive the motor, and the DC motor (which responds only to DC) develops no net torque! The system cannot get near a lock. There is no voltage available to bring the motor close to the lockup condition.

In communications loops the VCO (voltage controlled oscillator) has no inertia and can follow the beat frequency on an instantaneous basis. The inertialess communications VCO "accelerates" infinitely fast and thus reduces the difference between its output frequency (ω_{tach}) and the reference frequency, making a lockup possible.

In motor control loops the inertia of the motor and load prevents the motor from following the beat frequency on an instantaneous basis. Since the average value of the detector output is zero, no motion and no lockup occur. Only if the beat frequency is within the servo bandwidth (typically less than 100 Hz) is there a chance of lockup. This phase detector is clearly unsuitable for motor speed control.

PHASE DETECTORS—SWITCHING TYPES

Another common multiplying phase detector is implemented by using some form of switch. The "double balanced modulator," "synchronous detector,"

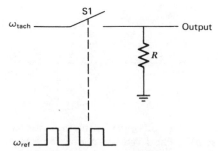

Figure 3-6 A basic switched phase detector. S1 is closed when ω_{ref} is high; S1 is open when ω_{ref} is low.

and "exclusive OR" are all essentially identical. Figure 3-6 shows a typical switched phase detector. Switch S1 is opened and closed at the rate of ω_{ref}. ω_{tach} is transmitted through the switch to load R. When switch S1 is closed, the tachometer waveform can be considered to be multiplied by unity. When switch S1 is open, the tachometer waveform can be considered to be multiplied by zero. This process is represented by the multiplication of the two signals.

Both waveforms are considered to be square waves, whose Fourier series may be written as

$$\text{Reference signal} = R_1 \cos(\omega_R t + \theta_{R_1}) + R_2 \cos(2\omega_R t + \theta_{R_2})$$
$$+ R_3 \cos(3\omega_R t + \theta_{R_3}) + \cdots \quad (3\text{-}9)$$

$$\text{Tachometer signal} = T_1 \cos(\omega_T t + \theta_{T_1}) + T_2 \cos(2\omega_T t + \theta_{T_2})$$
$$+ T_3 \cos(3\omega_T t + \theta_{T_3}) + \cdots \quad (3\text{-}10)$$

Each signal is an infinite series with an infinite number of harmonics because of the square wave nature of the signal. The exact values of the coefficients $R_1, R_2, R_3 \cdots$ and $T_1, T_2, T_3 \cdots$ are of secondary importance at this point. The multiplication process produces an output signal:

$$\text{Output} = \left(\sum_{k=1}^{\infty} R_k \cos(k\omega_R t + \theta_{R k}) \right) \times \left(\sum_{k=1}^{\infty} T_k \cos(k\omega_T t + \theta_{T k}) \right) \quad (3\text{-}11)$$

Each term of the tachometer signal multiplies each term of the reference signal. The resulting expression, *neglecting phase angles,* is

$$\text{Output} = R_1 T_1 \cos\omega_R t \cos\omega_T t + R_1 T_2 \cos\omega_R t \cos 2\omega_T t$$
$$+ R_1 T_3 \cos\omega_R t \cos 3\omega_T T + \cdots$$
$$+ R_2 T_1 \cos 2\omega_R t \cos \omega_T t + R_2 T_2 \cos 2\omega_R t \cos 2\omega_T t$$
$$+ R_2 T_3 \cos 2\omega_R t \cos 3\omega_T t + \cdots$$
$$+ R_3 T_1 \cos 3\omega_R t \cos \omega_T t + R_3 T_2 \cos 3\omega_R t \cos 2\omega_T t$$
$$+ R_3 T_3 \cos 3\omega_R t \cos 3\omega_T t + \cdots$$

$$\begin{array}{ccc} \bullet & \bullet & \bullet \\ \bullet & \bullet & \bullet \\ \bullet & \bullet & \bullet \end{array} \qquad (3\text{-}12)$$

An infinite collection of harmonic terms ensues. These may be organized in a representative symbolic fashion, which shows only the frequencies of the product pairs:

$$
\begin{array}{cccc}
\omega_R\,\omega_T & \omega_R\,2\omega_T & \omega_R\,3\omega_T & \omega_R\,4\omega_T \\
2\omega_R\,\omega_T & 2\omega_R\,2\omega_T & 2\omega_R\,3\omega_T & 2\omega_R\,4\omega_T \\
3\omega_R\,\omega_T & 3\omega_R\,2\omega_T & 3\omega_R\,3\omega_T & 3\omega_R\,4\omega_T \\
\vdots & \vdots & \vdots & \vdots
\end{array}
$$

where $\omega_R\,\omega_T$ represents the term $R_1 T_1 \cos\omega_R t \cos\omega_T t$. Each product term may be identified as a sum and difference term according to the identity of eq. 3-6. The chart may be rewritten:

$$
\begin{array}{cccc}
\omega_R + \omega_T & \omega_R + 2\omega_T & \omega_R + 3\omega_T & \omega_R + 4\omega_T \quad \cdots \\
\omega_R - \omega_T & \omega_R - 2\omega_T & \omega_R - 3\omega_T & \omega_R - 4\omega_T \\
\vdots & \vdots & \vdots & \vdots \\
2\omega_R + \omega_T & 2\omega_R + 2\omega_T & 2\omega_R + 3\omega_T & 2\omega_R + 4\omega_T \quad \cdots \\
2\omega_R - \omega_T & 2\omega_R - 2\omega_T & 2\omega_R - 3\omega_T & 2\omega_R - 4\omega_T \\
\vdots & \vdots & \vdots & \vdots \\
3\omega_R + \omega_T & 3\omega_R + 2\omega_T & 3\omega_R + 3\omega_T & 3\omega_R + 4\omega_T \quad \cdots \\
3\omega_R - \omega_T & 3\omega_R - 2\omega_T & 3\omega_R - 3\omega_T & 3\omega_R - 4\omega_T
\end{array}
$$

where $\omega_R + \omega_T$ represents the term $R_1 T_1 \cos\omega_R t \cos\omega_T t$. The sum of frequency terms may be ignored because they are generally far beyond the operating range of the loop. Only the difference terms are important here.

It can now be observed that this switching type of detector produces a DC output whenever $\omega_{\text{ref}} = \omega_{\text{tach}}$, just as in the case of the simple multiplier described earlier. However, a DC output is also produced where $2\omega_{\text{tach}} = \omega_{\text{ref}}$, or $\omega_{\text{tach}} = \omega_{\text{ref}}/2$. The motor can (and will) lock on a subharmonic! Terms such as $(9\omega_{\text{ref}} = 10\omega_{\text{tach}})$ also exist, implying that ω_{tach} can be $\frac{9}{10}\omega_{\text{ref}}$ and produce lockup! Since this is unacceptable for motor control applications, these phase detectors are limited to communications loops, where tuned circuits help to eliminate harmonic locking. As in the preceding multiplying scheme, no DC for acceleration exists.

PHASE-FREQUENCY DETECTORS

To overcome the problems of the multiplier-type phase detectors, a class of device often labeled "phase-frequency detector" was made available. It is exemplified by phase detector II in the 4046B CMOS chip, which is evaluated here. The general properties of the phase frequency detector take into account the three possible conditions for ω_{ref} and ω_{tach}:

Condition	Phase-Frequency Detector Output	Comment
1 $\omega_{tach} < \omega_{ref}$	1	Occurs during the initial acceleration to speed. The motor often starts from rest and accelerates maximally until lockup
2 $\omega_{tach} > \omega_{ref}$	0	Occurs when ω_{ref} is lowered to reduce motor speed. The motor decelerates maximally until lockup
3 $\omega_{ref} = \omega_{tach}$	Duty cycle proportional to phase difference	Occurs in the locked condition. The two signals differ only by their phase angle

Phase frequency detectors are digital circuits arranged to yield a logic 1 for condition 1, logic 0 for condition 2, and a duty cycle proportional to phase difference for condition 3. The equivalent output circuit is shown in Fig. 3-7. Whenever $\omega_{tach} < \omega_{ref}$, switch S1 is closed (S2 open), thereby delivering a logic 1 to the motor drive circuit. This connects V_{cc} to the output and provides voltage to accelerate the motor toward lockup speed. Whenever $\omega_{tach} > \omega_{ref}$, switch S2 is closed (S1 open), thereby producing a logic zero, usually 0 V output. This decelerates the motor toward zero speed and has the potential of producing lockup as ω_{tach} passes ω_{ref}.

At lockup and with a zero phase error between the reference and tachometer frequencies, S1 and S2 remain open and the output "floats." The charge stored on C provides a voltage to maintain motor speed.

When $\omega_{tach} = \omega_{ref}$ but ω_{tach} lags ω_{ref} by phase angle ϕ, the circuit action is as depicted in Fig. 3-8. The output is connected to V_{cc} (via S1) for

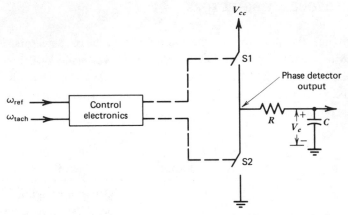

Figure 3-7 Simple equivalent circuit of a phase frequency detector.

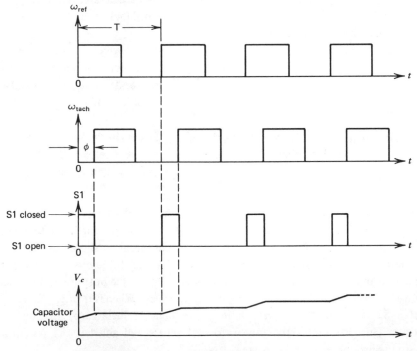

Figure 3-8 Waveforms for the simple phase frequency detector of Fig. 3-7. The top two waveforms show that the input reference frequency (ω_{ref}) leads the tachometer frequency (ω_{tach}) by $\Delta\phi$. The bottom two waveforms show the state of S1 and the resulting capacitor voltage. The capacitor is seen to be charging in an attempt to speed up the motor, thereby reducing $\Delta\phi$.

46

a time $(\phi/2\pi)T$, the fraction of time per cycle that represents the phase difference. The capacitor voltage increases, thereby speeding up the motor to allow the tachometer phase to align itself with the reference phase.

A similar action occurs when ω_{tach} leads ω_{ref}, except that S2 closes for a time $(\phi/2\pi)T$, decreasing the voltage on C. This slows the motor down and allows the tachometer phase to match the reference phase.

The circuit can be analyzed by considering the charging current of the capacitor due to a phase difference ϕ. The charging current (i) can be written as

$$i = \text{charging current} = \left(\frac{V_{cc} - v_c}{R}\right)\frac{\phi}{2\pi} \qquad (3\text{-}13)$$

where V_{cc} = supply voltage
v_c = capacitor voltage
R = resistance
ϕ = phase difference

S1 is closed for a fraction of the period proportional to the phase difference $\phi/2\pi$. The charging current during that period is simply the voltage across $R(= V_{cc} - v_c)$ divided by R (Ohm's law). Furthermore, the identical current charges capacitor C, so that

$$i = \text{capacitor charging current} = C\frac{dv_c}{dt} \qquad (3\text{-}14)$$

Equating the expressions for i, we have

$$\left(\frac{V_{cc} - v_c}{R}\right)\frac{\phi}{2\pi} = C\frac{dv_c}{dt} \qquad (3\text{-}15)$$

Rearranging:

$$\frac{dv_c}{dt} + v_c\frac{\phi}{2\pi RC} = V_{cc}\frac{\phi}{2\pi RC} \qquad (3\text{-}16)$$

Rewriting in transform notation:

$$sV_c + V_c\frac{\phi}{2\pi RC} = V_{cc}\frac{\phi}{2\pi RC} \qquad (3\text{-}17)$$

The solution of this equation is

$$V_c(s) = \frac{V_{cc}}{s\left(\frac{2\pi RC}{\phi}\right) + 1}$$

(3-18)

In the frequency domain:

$$V_c(j\omega) = \frac{V_{cc}}{j\omega\left(\frac{2\pi RC}{\phi}\right) + 1}$$

(3-19)

Assume the condition

$$\frac{\omega 2\pi RC}{\phi} \gg 1$$

(3-20)

This will be justified shortly. It follows that

$$V_c(j\omega) \approx \frac{V_{cc}\phi}{j\omega 2\pi RC}$$

(3-21)

and

$$\boxed{\frac{V_c}{\phi} = \frac{V_{cc}}{2\pi RC}\left(\frac{1}{j\omega}\right)}$$
transfer function of the phase detector

(3-22)

This equation represents the transfer function of the entire phase detector circuit. It remains to justify the assumption of eq. 3-20:

$$\frac{\omega 2\pi RC}{\phi} \gg 1$$

(3-20)

Rearranging the terms yields

$$\omega \gg \frac{\phi}{2\pi}\frac{1}{RC}$$

(3-23)

But $\phi/2\pi$ will always be less than unity. In other words,

$$\frac{\phi}{2\pi} \leqslant 1 \qquad (3\text{-}24)$$

A more stringent condition can then be substituted for eq. 3-23:

$$\omega \gg \frac{1}{RC} \qquad (3\text{-}25)$$

This says that the time constant of the circuit must be much larger than the reference (or tachometer) frequency (at lock).

It is interesting to evaluate the gain of this phase detector block for typical values. The 3 dB frequency of the RC filter is typically about 100 Hz [$(1/2\pi RC)$ = 100 Hz]. V_{cc} is typically 10 V. It follows that

$$\frac{V_c}{\phi} = \frac{V_{cc}}{2\pi RC}\frac{1}{j\omega} = \frac{10 \times 100}{j\omega} = \frac{1000}{j\omega} \quad (=60 \text{ dB at } \omega = 1 \text{ rad/sec}) \qquad (3\text{-}22)$$

This high gain is detrimental to the stable operation of the loop, and additional components are often added to reduce the phase gain.

PHASE-FREQUENCY DETECTOR—ITS MAIN PROBLEM

Another drawback of the phase frequency detector is its severe nonlinearity in the frequency domain (Fig. 3-9). For $\Delta\omega > 0$ (i.e., $\omega_{ref} > \omega_{tach}$) detector output is logic 1. For $\Delta\omega < 0$ (i.e., $\omega_{ref} < \omega_{tach}$) detector output is logic 0. At $\omega_{ref} = \omega_{tach}$ detector output is determined by the phase error between the two frequencies. If the equations are written in the frequency domain, the gain at lock is infinite!

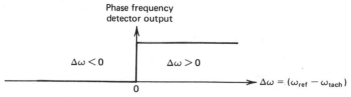

Figure 3-9 Phase frequency detector output versus frequency difference $\omega_{ref} - \omega_{tach}$. The phase frequency detector is nonlinear at $\omega_{ref} - \omega_{tach} = 0$.

What occurs in practice is that the motor/optics oscillates about the lock frequency until the *simultaneous* combination of *both* phase and frequency is right for lock, at which point the system snaps into lockup. If the shaft speed moves through the lockup conditions too rapidly, lockup becomes impossible. This results from a faster change in phase than the bandwidth of the system allows. It may be impossible to achieve lockup for a low bandwidth servo!

Furthermore, the extreme nonlinearity of the detector applies maximum drive to the motor, which exacerbates the lockup problem by sending the motor off to a "speeding start" toward lock. In other words, maximum acceleration or deceleration is applied immediately before lockup is supposed to occur. This creates large initial velocities, which must be removed before stable lock is achieved. *Note how the usage of ω as the independent variable has highlighted this problem.*

Although the phase frequency detector has been successfully used in motor control loops, the dual factors of excessive gain and nonlinearity in the frequency domain make it a less than optimum choice for error detection in motor control applications.

DIGITAL INTEGRATOR AND LOCK DETECTION

It can be shown that the configuration in Fig. 3-10 produces an output proportional to the time integral of the frequency error. The analog output of the D/A converter may be written as

$$\text{analog output} \approx k_i \int (f_{\text{ref}} - f_{\text{tach}})dt + C \qquad (3\text{-}26)$$

where k_i = integrator gain

C = constant

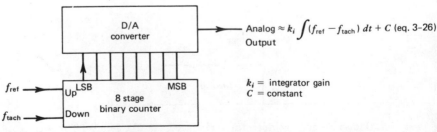

Figure 3-10 A digital counter and D/A (digital-to-analog) converter connected as an integrator of the difference between f_{ref} and f_{tach}.

A leading edge of f_{ref} counts the digital counter up. A leading edge of f_{tach} counts the digital counter down. If two edges occur simultaneously, a special clocking circuit inhibits both edges to avoid count ambiguities.

The output of the counter is connected to a D/A converter that produces an analog output directly proportional to the stored binary number. When the counter is "empty" (all bits = 0) the D/A converter output is $-\Delta V_{max}/2$. When the counter if "full" (all bits = 1) the D/A converter output is $+\Delta V_{max}/2$. (Typically $\Delta V_{max} = 10$ V.) A linear relationship between the binary count and output voltage exists for all numbers between these extremes. It can be written that

$$\Delta V_o = k_i(\Delta\text{count}) \tag{3.27}$$

with Δcount = change in counter content
$\quad\ \Delta V_o$ = change in D/A output voltage corresponding to Δcount

Furthermore

$$k_i = \frac{\Delta V_{max}}{\Delta\text{count}_{max}} \tag{3-28}$$

where ΔV_{max} = maximum D/A output voltage corresponding to Δcount_{max}
$\qquad\quad \Delta\text{count}_{max}$ = maximum count in the binary counter

For example, on assuming an 8 bit counter and 8 bit D/A, the maximum binary count is 11111111_2 ($= 256_{10}$). On assuming that ΔV_{max} is 10 V,

$$k_i = \frac{10\text{ V}}{256\text{ counts}} \cong 40\text{ mV/count}$$

The counter output rises approximately 40 mV on each up count or decreases 40 mV on each down count. k_i is typically adjustable between 0 and 40 mV/count.

When f_{tach} and f_{ref} are different, there will be a steady increase or decrease in counter contents. In a time period Δt the change in counter contents will be

$$(\Delta\text{ count}) = \Delta t(f_{ref} - f_{tach}) \tag{3-29}$$

Using eqs. 3-27 and 3-29:

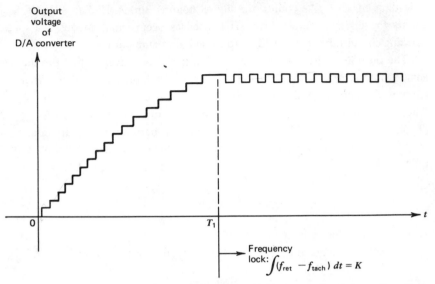

Figure 3-11 Typical output voltage of the D/A converter when the motor starts from rest and accelerates to the frequency locked condition.

$$\Delta V_o = k_i(\Delta \text{ count}) = k_i[\Delta t(f_{\text{ref}} - f_{\text{tach}})] \qquad (3\text{-}30)$$

Rearranging terms:

$$\frac{\Delta V_o}{\Delta t} = k_i(f_{\text{ref}} - f_{\text{tach}}) \qquad (3\text{-}31)$$

In the limit, for short time periods:

$$\lim_{\Delta t \to 0} \frac{\Delta V_o}{\Delta t} = \frac{dV_o}{dt} = k_i(f_{\text{ref}} - f_{\text{tach}}) \qquad (3\text{-}32)$$

Integrating both sides of the equation with respect to time:

$$V_o = k_i \int (f_{\text{ref}} - f_{\text{tach}}) dt + C \qquad (3\text{-}26)$$

Subject to the limit of resolution of the counter (8 bits in the example above), this circuit performs a true integration of the frequency error.

The usefulness of this integrating property lies in the fact that any error whatsoever between f_{ref} and f_{tach} produces an output voltage that continuously increases (or decreases) in time. Only exact correspondence between f_{ref} and f_{tach} can maintain a fixed error

Assuming that the motor starts from rest, a typical output waveform of the D/A converter is as shown in Fig. 3-11. Initially, the reference frequency drives the counter contents up very rapidly, because very few count down edges are produced by the slowly rotating motor. As the counter contents increase, an increasing error voltage is developed at the output of the D/A converter. This error voltage is transmitted to the motor, which speeds up, thereby causing more count down edges. As f_{tach} approaches f_{ref}, the counter contents increases only slowly. At time T_1, $f_{tach} = f_{ref}$ and the counter contents remains stationary. The stored count develops just enough output drive to the motor to keep $f_{ref} = f_{tach}$, thereby producing a "frequency locked" condition. Note that this is not true phaselock, because the phases are not detected.

It is interesting to observe the steady state output of the D/A converter during a frequency locked condition. Such output is shown in Fig. 3-12. A leading edge of f_{ref} causes an up count. A leading edge of f_{tach} causes a down count. The LSB (least significant bit) of the counter is continually changing, causing the D/A output to jump between V_1 and V_2. The duty cycle of LSB causes the DC output of the D/A converter to be

$$V_{avg} = \text{DC output of D/A converter} = V_1 + D(V_2 - V_1), \quad (D = \text{duty cycle})$$

The duty cycle "interpolates" between V_1 and V_2 and provides a continuously variable DC drive to the motor from the discrete output of the D/A converter.

Stated another way, the discrete jumps produced by the counter cannot, in general, provide the exact voltage necessary to satisfy motor operating requirements. In Fig. 3-12 V_1 is too low and V_2 is too high. The duty cycle waveform effectively fills in the gap between V_1 and V_2. This interpolation gives the

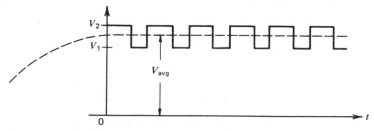

Figure 3-12 Steady state output of D/A converter at frequency lock.

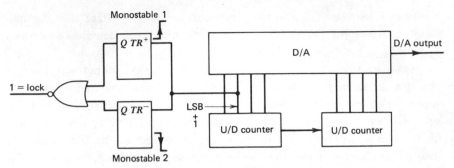

Figure 3-13 The LSB + 1 can be used to make a lock indicator. At lock the LSB + 1 (and all higher bits) is stable. It is either a 1 or a 0, and is not changing. The monostables will not be triggered, and a 1 appears at the output of the NOR gate.

exact output necessary for proper motor speed and torque.

From the foregoing discussion it can be concluded that at frequency lock only the LSB changes. All other counter outputs remain constant at either logic 1 or logic 0. This suggests that a lock indication can be obtained from the fact that the counter content (above the LSB) is static. If the counter content were changing, by definition there would be more edges from f_{ref} than from f_{tach} (or vice versa).

Monitoring the output of the LSB + 1 (the next least significant bit) and determining whether it is changing or stationary can be the basis of an effective lock indicator. Such a system is shown in Fig. 3-13. Monostable 1 triggers on positive-going edges. Monostable 2 triggers on negative-going edges. If LSB + 1 is stationary, no triggering of the monostables occurs, and the NOR gate output

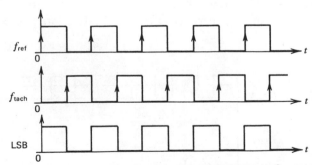

Figure 3-14 When f_{ref} and f_{tach} differ in phase by 180°, the LSB of the counter exhibits a 50 % duty cycle.

is logic 1. This indicates frequency lock. When the LSB + 1 changes, one of the monostables will be triggered, causing a logical 0 at the NOR output. This indicates an "out of lock" condition. The period of the monostables is often chosen to be approximately the time for 100 cycles of the lowest f_{ref} being used. Using this choice, f_{ref} and f_{tach} must be within 1% of each other to cause a lock indication. More exact indication can be obtained by making the monostable period longer, but this is rarely necessary. If a steady state can be reached at all, it very quickly follows the 1% frequency difference condition. Consequently, the criterion above for setting the period of the monostables works well in practice.

It has been shown how the counter D/A converter produces the frequency locked condition. In Chapter 5 it is shown how to obtain a true phaselock by forcing the duty cycle of the LSB to 50% by auxiliary means. If the LSB is always at 50% duty cycle, a phaselocked condition must exist, because the leading edges of f_{ref} and f_{tach} are always identically "spaced" (Fig. 3-14).

4

Velocity Servos Using the Optical Tachometer

This chapter assumes that the reader is familiar with control system theory, particularly with second order systems. Review material on control theory is given in Appendixes A, B, and C and in the references. Equations that completely characterize the velocity servo (which uses the optical tachometer as the feedback transducer) are derived here. These equations are used in a flowchart to provide an organized design process.

The analysis of the velocity servo can be divided into two parts. The first part is the dynamic analysis, which describes the dynamic characteristics of the velocity servo. The second part is the static analysis, which describes static performance, specifically the regulation under torque load. Servo parameters must be chosen to satisfy both static and dynamic criteria simultaneously. The dynamic analysis is considered first.

DYNAMIC ANALYSIS

A block diagram of the velocity servo is shown in Fig. 4-1 and can be briefly described as follows. The reference frequency ω_{ref} is fed to an F/V (frequency-to-voltage) converter. Similarly, the output of the optical tachometer ω_{tach} is fed to an F/V converter. (This may be the four quadrant velocity sensing circuit described in Chapter 3.) An error voltage proportional to $\omega_{ref} - \omega_{tach}$ is developed and transmitted to error amplifier G_1, then to transconductance amplifier A_1, and then to the motor, which converts the resulting current to shaft rotation. Optical disc N provides feedback signal ω_{tach}.

The servo loop has two explicit poles: one from the time constant $(= \tau_M)$ of the filter circuit associated with the monostable and the other from the inertia (J). In addition, a phase shift giving the effect of a third (implicit) pole is caused by the discrete nature of the optical tachometer output. The optical tachometer delivers an output frequency of f_{tach}, and the information arrives every $1/f_{tach}$ sec. An effective time delay of $1/f_{tach}$ is introduced by the discontinuous nature of the velocity feedback signal. Since only the leading edges of the tachometer waveform contain useful rate information, the time between edges $(= 1/f_{tach})$ is "dead time" which is effectively a delay. To avoid or minimize this delay, f_{tach} can be made 10 times higher than system bandwidth. The phase shift (i.e., delay) caused by the "sampled data" process is thereby rendered negligible. In other words, if $1/f_{tach}$ is 10 times smaller than the time constant of the system, sampling delay is negligible and the servo is considered to have only two poles.

These thoughts on sampling delay may be quantified. The requirement for negligible phase shift from sampling is

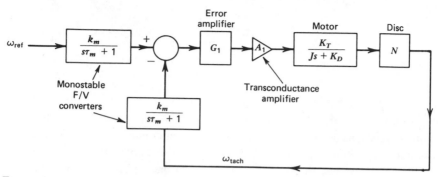

Figure 4-1 Velocity servo block diagram suitable for analysis. k_m = monostable gain (V/rad/sec); τ_M = monostable filter time constant (sec); G_1 = amplifier gain (V/V); A_1 = transconductance (A/V); K_T = motor torque constant (oz-in/A); J = inertia (oz-in-sec^2); K_D = damping constant (oz-in/1000 rpm); N = disc line density.

$$\omega_{\text{tach min}} \geqslant 10 \times \text{bandwidth} \qquad (4\text{-}1)$$

where bandwidth has units of radians/second (rad/sec) and $\omega_{\text{tach min}}$ = minimum allowable tachometer frequency that avoids "sampled data" phase shift. For a critically damped second order system it is known that the "natural frequency" = bandwidth. For other conditions of damping, the equation "natural frequency" = bandwidth is only an approximation, albeit a good one. Using eq. 4-1 and substituting ω_N for bandwidth yields eq. 4-2:

$$\omega_{\text{tach min}} \geqslant 10\,\omega_N \qquad (4\text{-}2)$$

where ω_N = natural frequency of a second order system. $\omega_{\text{tach min}}$ is directly related to minimum rpm by eq. 2-1, which is repeated below for convenient reference.

$$\omega_{\text{tach min}} = \frac{\pi}{30} N\,\text{rpm}_{\text{min}} \qquad (2\text{-}1)$$

where N = disc density

rpm$_{\text{min}}$ = minimum allowable motor speed for negligible sampling delay

Setting the expressions for $\omega_{\text{tach min}}$ equal in eqs. 4-2 and 2-1, and solving for N yields:

$$N \geqslant \frac{300\, \omega_N}{\pi\, \mathrm{rpm}_{\min}} \qquad (4\text{-}3)$$

This equation is fundamental to all that follows and determines the limits of performance on both velocity servos and PLL servos. It provides a very important auxiliary relationship between bandwidth (ω_N), minimum rpm, and disc density. This equation will be repeatedly used in much of the work to follow.

Figure 4-1 shows that both the tachometer and reference frequencies are converted to voltages by their respective monostable F/V converters. The difference between these voltages is amplified by G_1, converted to current by A_1, transformed to shaft rotation by the motor, and fed back to the summing junction through the optical disc (N lines). Using standard techniques, the F/V converter blocks can be moved beyond the summing junction, and the block diagram can be redrawn in a condensed fashion as in Fig. 4-2.

From the block diagram the open loop transfer function can be written as

$$G(s) = \frac{k_m G_1 A_1 K_T N}{(s\tau_M + 1)(Js + K_D)} \qquad (4\text{-}4)$$

where

$$k_m = \frac{V_{cc}}{\omega_{\text{tach max}}} \qquad (4\text{-}4a)$$

and

$$\omega_{\text{tach max}} = \frac{\pi}{30} N\, \mathrm{rpm}_{\max} \qquad (4\text{-}4b)$$

On factoring out K_D, and defining K and τ_J as shown, it follows that

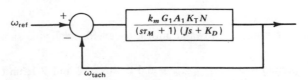

Figure 4-2 Simplified velocity servo block diagram.

$$G(s) = \frac{K}{(s\tau_M + 1)(s\tau_J + 1)} \tag{4-5}$$

where

$$\tau_J = \frac{J}{K_D} \tag{4-5a}$$

$$K = \frac{k_m G_1 A_1 K_T N}{K_D} \tag{4-5b}$$

The closed loop transfer function may be written as

$$\frac{G(s)}{G(s) + 1} = \frac{K/(s\tau_M + 1)(Js + K_D)}{[K/(s\tau_M + 1)(Js + K_D)] + 1}$$

$$= \frac{K}{s^2 \tau_M \tau_J + s(\tau_M + \tau_J) + (K + 1)} \tag{4-6}$$

The characteristic equation is

$$s^2 \tau_M \tau_J + s(\tau_M + \tau_J) + K + 1 = 0 \tag{4-7}$$

This is rearranged as

$$s^2 + s\frac{(\tau_M + \tau_J)}{\tau_M \tau_J} + \frac{K + 1}{\tau_M \tau_J} = 0 \tag{4-8}$$

By comparison with the standard form of the second order equation* it may be written that

$$\omega_N{}^2 = \frac{K + 1}{\tau_M \tau_J} \tag{4-9}$$

where ω_N = natural frequency and

*The standard form of the second order equation is $s^2 + 2\zeta\omega_N s + \omega_N{}^2 = 0$ where ω_N = natural frequency $\approx$ bandwidth and ζ = damping ratio.

$$2\zeta\omega_N = \frac{\tau_M + \tau_J}{\tau_M \tau_J} = \frac{1}{\tau_M}\left(1 + \frac{\tau_M}{\tau_J}\right) \tag{4-10}$$

where ζ = damping ratio. In most cases of practical interest $\tau_M/\tau_J \ll 1$. With this assumption, eq. 4-10 becomes

$$2\zeta\omega_N = \frac{1}{\tau_M} \tag{4-11}$$

If eqs. 4-9 and 4-11 are solved simultaneously for τ_M and ω_N, the results are

$$\tau_M = \frac{\tau_J}{4\zeta^2(K + 1)} \tag{4-12}$$

and

$$\omega_N = \frac{2\zeta(K + 1)}{\tau_J} \tag{4-13}$$

These equations are used later to help select the proper values of τ_M, ω_N, and ζ.

STATIC ANALYSIS

The static analysis is based on the block diagram shown in Fig. 4-3. This is a direct offshoot of Fig. 4-1, except that only static conditions are considered. The motor has been split into two blocks. The first block, labeled K_T, converts the current from the amplifier A_1 into torque. This torque is the load torque T_L, friction torque T_f, and torque to overcome damping T_D. The second block, labeled K_D, converts damping torque T_D to velocity. This "split" of the torque components may be clarified as follows. According to Newton's law, equal and opposite reaction torques, exactly equal to T_L and T_f, are developed by the load and friction respectively. A specific value of torque is required to overcome damping and maintain motor speed. T_D is the torque that must be supplied to hold the motor at the required speed. Stated another way, T_L and T_f are balanced out by the load and friction torques respectively. T_D is balanced out

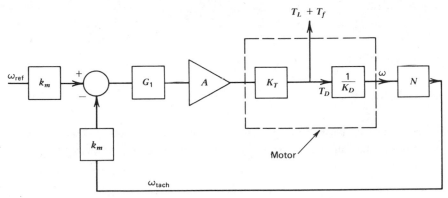

Figure 4-3 Block diagram for the static analysis.

by the motor damping when the motor reaches the required speed. The static balance equation may be written as

$$(\omega_{\text{ref}} k_m - \omega_{\text{shaft}} N k_m) G_1 A_1 K_T = T_L + T_f + T_D \qquad (4\text{-}14)$$

where T_L = load torque
            T_f = friction torque
            T_D = damping torque
      ω_{shaft} = shaft speed (rad/sec)

Differentiating both sides with respect to T_L:

$$-\frac{\partial \omega_{\text{shaft}}}{\partial T_L} N k_m G_1 A_1 K_T = 1 \qquad (4\text{-}15)$$

Solving for the differential term:

$$\frac{\partial \omega_{\text{shaft}}}{\partial T_L} = \frac{-1}{N k_m G_1 A_1 K_T} \qquad (4\text{-}16)$$

Converting to rpm and defining R as "regulation," we have

$$\frac{\partial n}{\partial T_L} = R = -\frac{30/\pi}{N k_m G_1 A_1 K_T} \qquad (4\text{-}17)$$

where n = rpm

$\quad$ R = regulation (rpm/oz-in).

This equation will be used to select $G_1 A_1$.

COMBINING THE STATIC AND DYNAMIC ANALYSES

To develop a design procedure, it is helpful to combine some of the foregoing equations in the following manner. Solve eq. 4-5b for $G_1 A_1$, which gives

$$G_1 A_1 = \frac{KK_D}{k_m K_T N} \tag{4-18}$$

Substitute eq. 4-18 into eq. 4-17 to obtain

$$R = - \frac{30/\pi}{K_T k_m N \left(\dfrac{KK_D}{k_m K_T N} \right)} = \frac{30}{\pi K K_D} \tag{4-19}$$

Solve eq. 4-13 for K to obtain

$$\frac{\omega_N \tau_J}{2\zeta} - 1 = K \tag{4-20}$$

Substitute eq. 4-20 into eq. 4-19 to yield

$$R = - \frac{30}{\pi K_D (\omega_N \tau_J / 2\zeta - 1)} \tag{4-21}$$

It is helpful to solve eq. 4-21 for ω_N. This solution is

$$\omega_N = \frac{2\zeta}{\tau_J} \left(-\frac{30}{\pi} \frac{1}{R K_D} + 1 \right) \tag{4-22}$$

Note that R is usually a negative number. Equations 4-21 and 4-22 (which express one relationship in two forms) are the key to proper design of velocity servos. Once the relative stability ζ, the damping K_D, and inertia J are chosen, only one of the two remaining variables (ω_N and R) may be chosen independently. In addition, eq. 4-3 sets further limits on ω_N, N, and rpm min.

One other limitation can often be imposed. N must often be less than some N_{max}. Typically, as line density N increases, so does cost. Furthermore, for a given disc diameter, there is an ultimate limit on N because of the wavelength of light. The spaces between lines certainly cannot be narrower than the wavelength of light! This is an extreme case, and for many practical purposes N is limited to 5000 lines or less (on a 2-½ in. diameter disc). In general, it may be written that

$$\boxed{N \leqslant N_{max}} \qquad\qquad (4\text{-}23)$$

All the design equations are now available and a design procedure can be developed. The problem statement rarely contains exactly the right information for proper design of the velocity servo. It is assumed here that the motor has been chosen to provide enough torque at the requisite speed and temperature. Choosing a proper motor is outside the scope of the present discussion, and substantial literature is available on that subject. Once the motor is chosen, it should be possible to obtain numbers for K_T, K_D, and the motor inertia. Furthermore, the load inertia should be known, or at least estimated, and the total inertial load J can be established.

In addition V_{cc}, the logic supply voltage, should be specified. ζ, the damping ratio should be selected ($\zeta = 1$ for critical damping). Also, rpm$_{max}$ must be specified in the problem statement or by the designer. From this discussion the "known" variables are considered to be: J, K_T, K_D, N_{max}, V_{cc}, ζ, and rpm$_{max}$.

Typically, the problem statement contains some information about disc density N, bandwidth (= ω_N), regulation R, and/or low end speed rpm$_{min}$. As pointed out earlier, these variables cannot be specified independently, but must satisfy eqs. 4-21, 4-3, and 4-23. It is the task of the designer to determine the four variables so that all constraints are met. This is the heart of the design problem. Once this is done, the remaining parameters (τ_M and $G_1 A_1$) are uniquely determined.

Chart 4-1 shows all the possible ways in which the problem can reach the designer. It may be overspecified, underspecified, or properly defined. For each

CHART 4-1 SPECIFIED PARAMETERS AND ASSOCIATED TESTS TO ENSURE REALIZABILITY

Case	Specified Parameters	Required Tests
1	$N, \omega_N, R, \text{rpm}_{\min}$	1. Substitute the specified ω_N into eq. 4-21 and find a value for R. Is this R less than or equal to the specified R? If no, specification can not be met. If yes, specification is valid
		2. Substitute ω_N and $\text{rpm}_{\min}$ into eq. 4-3 and find N. Is the specified N greater than N calculated from eq. 4-3? If yes, specification is valid. If no, specification can not be met
2	N, ω_N, R	1. Do step 1 of case 1
3	$\omega_N, R, \text{rpm}_{\min}$	1. Do step 1 of case 1
		2. Substitute ω_N and $\text{rpm}_{\min}$ into eq. 4-3 and find N. Is $N \leqslant N_{\max}$? If yes, specification is valid. If no, specification can not be met
4	$N, \omega_N, \text{rpm}_{\min}$	1. Do step 2 of case 1
5	$N, R, \text{rpm}_{\min}$	1. Substitute the specified value of R into eq. 4-22 and find ω_N; then
		2. Do step 2 of case 1
6	N, ω_N	
7	N, R	No design problem. Specification can always be met
8	$N, \text{rpm}_{\min}$	
9	$\omega_N, \text{rpm}_{\min}$	1. Do step 2 of case 3
10	ω_N, R	1. Do step 1 of case 1
11	$R, \text{rpm}_{\min}$	1. Do step 1 of case 5
		2. Do step 2 of case 3
12	N	1. $N \leqslant N_{\max}$?

**CHART 4-1 SPECIFIED PARAMETERS
AND ASSOCIATED TESTS TO ENSURE REALIZABILITY
(continued)**

Case	Specified Parameters	Required Tests
13	ω_N	
14	R	Underspecified
15	rpm_{min}	

possible specification an associated test is listed that must be performed to determine whether or not the specification is contradictory within itself. The "specified parameter" should always be a "worst case" value (largest acceptable N, lowest acceptable ω_N, largest R, and largest rpm_{min}).

Once it is established that no contradiction occurs (or the contradiction is resolved), Chart 4-2 can be used to calculate the remaining variables.

CHART 4-2 FLOWCHART FOR DESIGNING VELOCITY SERVOS

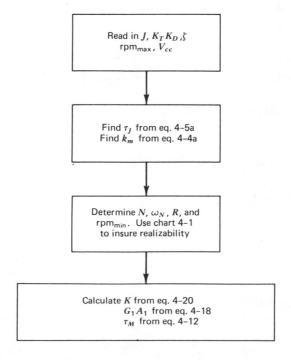

Design Example

A PMI U12M4H* with 5000 line optics is to be in a velocity servos that has a speed range from 30 to 3000 rpm. The inertial load consists of motor and disc only. The supply voltage is +15 V. Design the velocity servo.

From the problem statement:

$$N = 5000$$

$$\text{rpm}_{min} = \text{LRPM} = 30$$

$$\text{rpm}_{max} = \text{HRPM} = 3000$$

$$V_{cc} = 15 \text{ V DC}$$

From the data sheet for a PMI U12M4H:

$$K_T = 27 \text{ A/oz-in.}$$

$$K_D = 4.7 \text{ oz-in./1000 rpm}$$

$$J_{motor} = 0.02 \text{ oz-in-sec}^2$$

The inertia of the optics disc is known to be

$$J_{optics} = 0.002 \text{ oz-in-sec}^2$$

The total inertia is the sum $(J_{motor} + J_{optics}) = 0.02 + 0.002 = 0.022$ oz-in. sec^2.

The design may proceed as follows:

Step 1

Use Chart 4-1 to establish the validity of the specification of the four variables N, ω_N, R, and rpm$_{min}$. Only N and rpm$_{min}$ are specified. This corresponds to case 8. There is no conflict in the specification.

Step 2

Use Chart 4-2 as follows:

*Note: See Appendix F for Motor Data Sheet.

(a) Select $\zeta = 1$ (for critical damping).

(b) Find τ_J from eq. 4-5a as

$$\tau_J = \frac{J}{K_D} = \frac{0.022 \text{ oz-in. sec}^2}{4.7 \text{ oz-in.}/1000 \text{ rpm} \times 60 \text{ rpm}/1 \text{ rps} \times 1 \text{ rps}/2\pi \text{ rad/sec}}$$

$$= 0.49 \text{ sec}$$

Note that conversion to radian measure (for K_D) is necessary.

(c) Find k_m from eq. 4-4a:

$$k_m = \frac{V_{cc}}{\omega_{\text{tach max}}} = \frac{15}{(N \text{ rpm}_{\text{max}} \times 2\pi/60)} = \frac{15}{(5000 \times 3000 \times 2\pi/60)}$$

$$= 9.55 \times 10^{-6} \text{ V/rad/sec}$$

Step 3

Use eq. 4-21 or 4-22 to choose ω_N and R. If it is important to have a "tight" velocity servo (i.e., small R), use eq. 4-22. This permits the selection of a desired value for R and calculates the resultant ω_N. If it is important to have a wide bandwidth servo, use eq. 4-21. Choose a value for ω_N and determine the resultant R. In the present case, it is instructive to calculate a range of values for ω_N for a given range of R. With $R = -1$ rpm/oz-in., ω_N is calculated from eq. 4-22.

$$\omega_N = \frac{2\zeta}{\tau_J}\left(-\frac{30}{\pi}\frac{1}{RK_D} + 1\right)$$

$$= \frac{2 \times 1}{0.49}\left(\frac{-30}{\pi}\frac{1}{-1 \text{ rpm/oz-in.} \times 4.7 \text{ oz-in.}/1000 \text{ rpm}} + 1\right)$$

$$\omega_N = 8297 \text{ rad/sec}$$

$$f_N = \frac{\omega_N}{2\pi} = 1320 \text{ Hz}$$

A number of such calculations can be quickly made on a programmable calculator and the following chart established:

Regulation R	f_N Bandwidth
$-\frac{1}{2}$ rpm/oz-in.	2640 Hz
-1 rpm/oz-in.	1320 Hz
-2 rpm/oz-in.	660 Hz
-3 rpm/oz-in.	440 Hz
-4 rpm/oz-in.	330 Hz
-5 rpm/oz-in.	265 Hz

The regulation of a PMI U12M4H motor running from a simple regulated DC voltage source (i.e., no feedback) is listed on the data sheet as 2.2 rpm/oz-in. Clearly it makes no sense to let R be more than 2 because such a servo offers no advantage over a simple regulated power supply. Consequently, R should be chosen as 2 or less (in absolute value). If $R = \frac{1}{2}$, the bandwidth is somewhat wide and may result in oscillation because other poles can become significant. Therefore, a regulation of unity is chosen. It offers improved regulation over the intrinsic motor without extremes of bandwidth.

K is calculated from eq. 4-20:

$$K = \frac{\omega_N \tau_J}{2\zeta} - 1 = \frac{(1320 \times 2\pi) \times 0.49}{2 \times 1} - 1 = 2030$$

$G_1 A_1$ if found from eq. 4-18:

$$G_1 A_1 = \frac{K K_D}{k_m K_T N}$$

$$= \frac{2030 \times (4.7 \text{ oz-in.}/1000 \text{ rpm} \times 60 \text{ rpm}/1 \text{ rps} \times 1 \text{ rps}/2\pi \text{ rad/sec})}{9.55 \times 10^{-6} \text{ V/rad/sec} \times 27 \text{ oz-in.}/A \times 5000}$$

$$= 70.66 \approx 70$$

τ_M is determined from eq. 4-12:

$$\tau_M = \frac{\tau_J}{4\zeta^2(K + 1)} = \frac{0.49}{4 \times 1^2(2031)} \approx 60 \text{ } \mu\text{sec}$$

From these results, the amplifiers G_1 and A_1 must give a total gain of about 70. The monostable time constant is selected to give a cutoff frequency of

$$\omega_M = \frac{1}{\tau_M} = 16{,}666 \text{ rad/sec}$$

$$f_M \cong 2650 \text{ Hz}$$

As a check, the minimum tachometer frequency is calculated as

$$f_{\text{tach min}} = \frac{N \text{ rpm}_{\text{min}}}{60} = \frac{5000 \times 30}{60} = 2500 \text{ Hz}$$

Only about 3 dB of filtering is present at lowest rpm (recall that $2\pi/\tau_M = 2650$ Hz). The motor will probably emit some sound ($= 2500$ Hz) at the lowest rpm.

5

The Frequency Lock Loop

To achieve zero speed error, a pure integration must be inserted into the velocity loop. The velocity loop is thereby converted to a frequency locked loop in which a cycle-for-cycle correspondence between ω_{ref} and ω_{tach} exists. The pure integration is produced by the digital integrator described in Chapter 3. The block diagram for the frequency locked loop is given in Fig. 5-1.

The block diagram shows two parallel error processing paths. The "upper" path carries the velocity error signal $\omega_{ref} - \omega_{tach}$. The "lower" path carries the integrated velocity error signal $(\int(\omega_{ref} - \omega_{tach})dt)$. Laplace notation is used to express the integration (i.e., $1/s$). The "amounts" of velocity error signal, and integrated velocity error signal, are controlled by G_1 and k_i respectively.

It is the integration that forces a zero steady state error between the tachometer and reference frequencies. If the frequencies are at all different, the counter-D/A circuit produces a steadily increasing (or decreasing) error voltage, which drives the motor toward equilibrium. Only when the two frequencies are absolutely identical does the output of the integrator cease to change. At that point the output of the integrator makes up the difference between the velocity error signal from G_1 and the voltage necessary to drive the motor at the exact speed for ω_{ref} to equal ω_{tach}. To clarify this point, consider the analog system in Fig. 5-2, which shows a general servo loop with a pure integrator.

From the block diagram (assuming a constant error):

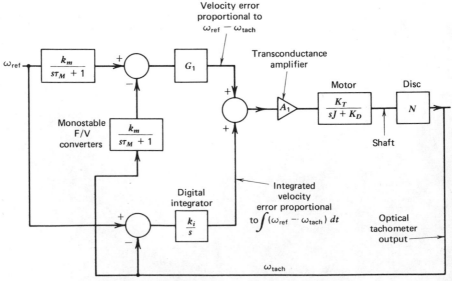

Figure 5-1 Block diagram for the static analysis.

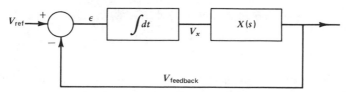

Figure 5-2 Block diagram of a servo that includes a pure integration in the forward path.

$$V_x = \int \epsilon \, dt = \epsilon t$$

where $\epsilon = V_{ref} - V_{feedback}$. If $\epsilon \neq 0$, V_x increases (or decreases) indefinitely. Only if $\epsilon = 0$ can a steady state be reached. When $\epsilon = 0$, $V_{ref} = V_{feedback}$ exactly! The insertion of a pure integration reduces the error to zero. This is an example of a type 1 servo. The frequency locked loop is another type 1 servo, since it contains one pure integration in the signal path. Type 1 servos produce zero error, but are somewhat difficult to stabilize. If both integral and proportional signals are used instead of just the integrator, the steady state condition still has zero error, but stability is easier to achieve than with only an integrator in the forward apth.

Returning to Fig. 5-1, let us simplify it by moving the k_m, G_1, and k_i blocks beyond the second summing junction. Fig. 5-3 results, from which the open loop transfer function, eq. 5-1, is written. The feedback path is arranged to have a transfer function of unity ($H(s) = 1$):

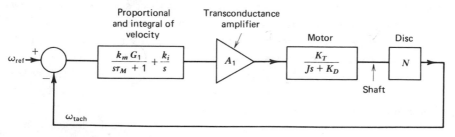

Figure 5-3 Simplified block diagram of the frequency locked loop.

$$G(s) = \left(\frac{G_1 k_m}{s\tau_M + 1} + \frac{k_i}{s} \right) A_1 K_T N \left(\frac{1}{Js + K_D} \right) \quad (H(s) = 1) \quad (5\text{-}1)$$

By algebra:

$$G(s) = \left(\frac{s(G_1 k_m / k_i + k_i / s) + 1}{s(s\tau_M + 1)(s\tau_J + 1)} \right) \left(\frac{A_1 K_T N k_i}{K_D} \right) \qquad \left(\tau_J = \frac{J}{K_D} \right) \qquad \text{(5-2)}$$

Equation 5-2 represents the open loop transfer function of the frequency locked loop. The frequency locked loop has two problems associated with it. First, the open loop gain $(A_1 K_T N k_i / K_D)$ is primarily controlled by the gain of the integrator k_i. It is well known that increasing the open loop gain of a servo tends to make it more oscillatory and closer to instability. Consequently, only limited amounts of integration can be used.

It may be theorized that even "very small" amounts of integration will reduce steady state velocity error to zero, and this is indeed the case. However, it turns out in practice that the counter D/A converter would have to count many cycles if only a small amount of integration gain "k_i" were used. This may be better understood by considering the action of the D/A integrator. When a torque distrubance is applied, the shaft slows down for an instant. ω_{tach} becomes slightly less than ω_{ref} for a very short time. The D/A integrator reacts to this frequency difference by counting up some number of cycles, thus increasing the drive to the motor and bringing ω_{tach} back to exact synchronism with ω_{ref}. The counter has counted up some number of cycles, but how many cycles? The number of cycles the counter has to move depends on gain k_i. If k_i is large, only one or two cycles (i.e., one or two counts) may be necessary to restore synchronous speed. If k_i is very high, perhaps only the duty cycle of the LSB of the counter need increase.

On the other hand, if k_i is low, it may take 10, 20, or even 100 cycles to count the counter up (or down) to restore synchronous speed. If k_i is too low, the counter may run out of range (i.e., greater than 256 counts for an 8 bit counter), and synchronous speed cannot be restored! By increasing the gain k_i, the counter has to count up less cycles to maintain speed. However, the system becomes increasingly difficult to stabilize. Clearly, the "stiffness" of this system as it now stands is not well controlled. The limit on k_i does not permit high gain. This is one problem with the frequency locked loop.

Another, probably more severe, problem of the frequency locked loop is that the D/A integrator does not operate well in the region of transition from one digital level to the next. This results from the circuit arrangement that rejects coincident edges of tachometer and reference signals. To avoid ambiguous counting, the D/A integrator must be operated to avoid the transition region between digital levels.

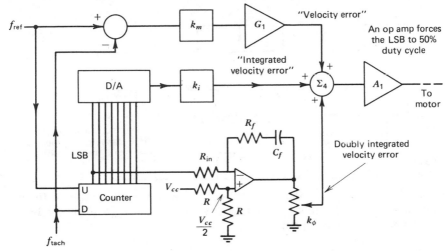

Figure 5-4 The op amp is connected to compare the DC component of the counter LSB output to $V_{cc}/2$. Any difference is integrated by the circuit and causes a change in speed. Only when the DC component of the LSB output is exactly $V_{cc}/2$ (i.e., 50 % duty cycle) will the op amp output cease to change. *Note:* Op amp transfer function is $V_o/V_{in} = (s\tau_1 + 1)/s\tau_2$ where $\tau_1 = C_f R_f$, $\tau_2 = C_f R_{in}$, V_{in} = input to R_{in}, and V_o = op amp output voltage.

PHASE LOCKING

To overcome difficulties just described, an additional integrator op amp circuit is added to operate on the LSB of the counter. The circuit/block diagram is shown in Fig. 5-4. Two signal paths have already been described. The op amp circuit forms a third signal path, which starts at the LSB of the counter and ends at the summing junction Σ_4. It will now be analyzed.

At lockup, the output of the LSB has a DC component directly proportional to the phase difference between the tachometer and reference frequencies. This is shown in Fig. 5-5 where f_{ref} counts the LSB up and f_{tach} counts the LSB down. The time difference between rising edges of f_{ref} and f_{tach} is defined as the phase difference between f_{ref} and f_{tach}. This is due to the normal action of the U/D counter, and is described in Chapter 3.

Since the op amp circuit is capacitively coupled to the output through C_f, the DC component of the op amp output cannot directly influence the input circuit. The op amp output reaches a steady state only when the DC component

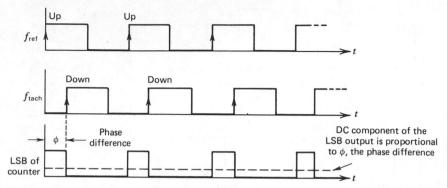

Figure 5-5 The LSB of the counter is "1" for a time proportional to the phase difference between f_{ref} and f_{tach}. When f_{ref} and f_{tach} differ by 180°, the LSB has a 50 % duty cycle.

of the duty cycle waveform is exactly $V_{cc}/2$, thus matching the DC voltage on the noninverting input of the op amp. The duty cycle of the LSB must be exactly 50% for this to occur. The op amp output is connected to the drive system (A_1) and helps to drive the motor so that the duty cycle of the LSB is forced to 50%. Only when the duty cycle attains 50% will the op amp output cease to change. Clearly this avoids operation near the transition region of the counter and solves at least one of the problems described above.

The fact that the additional op amp circuit forces 50% duty cycle on the LSB of the counter causes an 180° relationship between the tachometer and reference frequencies. This is easy to see from Fig. 5-5, which shows an approximate 25% duty cycle for the LSB. When the duty cycle increases to 50%, the phases of the tachometer and reference frequencies are 180° apart. The two signals are actually phaselocked at 180°.

The inclusion of the op amp circuit changes the servo block diagram to that of Fig. 5-6, which includes the third signal path. This block diagram represents the final configuration of the motor control PLL. It will be analyzed in the remaining portion of this chapter.

The analysis of the complete PLL system will result in a set of equations that can be used to determine all loop parameters. These equations can be solved by using a computer (or programmable calculator) to reduce the design of the PLL to an organized, routine process. The analysis begins with the block diagram of Fig. 5-6. All loop components are shown, and a general set of equations is derived. To begin, the open loop transfer function is written directly by observation of the block diagram:

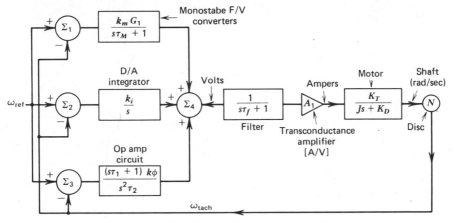

Figure 5-6 PLL block diagram suitable for analysis.

$$G(s) = \left[\frac{k_m G_1}{s\tau_M + 1} + \frac{k_i}{s} + \frac{s\tau_1 + 1}{s^2\tau_2} k_\phi \right] \frac{A_1}{s\tau_f + 1} \frac{NK_T}{Js + K_D} \quad (H(s) = 1) \quad (5\text{-}3)$$

where k_m = monostable "gain"

G_1 = error amplifier gain (dimensionless)

τ_M = monostable filter time constant (sec)

k_i = integration "gain" (V/count)

τ_1, τ_2 = op amp time constants (sec)

k_ϕ = phase gain (dimensionless)

A_1 = transconductance (A/V)

τ_f = carrier filter time constant (sec)

N = number of lines on the disc (dimensionless)

K_T = motor torque constant (oz-in./A)

J = total inertia (oz-in-sec^2)

K_D = damping constant (oz-in./1000 rpm)

When the term containing the sum is put over a common denominator, and τ_J is defined as in Chapter 4, eq. 5-3 can be rewritten as

$$G(s) = \left[\frac{k_m G_1 s^2 + sk_i(s\tau_M + 1) + (s\tau_1 + 1)(s\tau_M + 1)\dfrac{k_\phi}{\tau_2}}{s^2(s\tau_M + 1)} \right] \frac{A_1 K_T N/K_D}{(s\tau_f + 1)(s\tau_J + 1)}$$

$$(5\text{-}4)$$

where

$$\tau_J = \frac{J}{K_D} \tag{5-5}$$

and

$$\omega_J = \frac{1}{\tau_J} = \frac{K_D}{J} \tag{5-5a}$$

By collecting like powers of s in the numerator

$$G(s) = \left[\frac{s^2 \left(k_m G_1 + k_i \tau_M + \tau_M \tau_1 \frac{k_\phi}{\tau_2} \right) + s \left(k_i + (\tau_1 + \tau_M) \frac{k_\phi}{\tau_2} \right) + \frac{k_\phi}{\tau_2}}{s^2 (s\tau_M + 1)(s\tau_f + 1)(s\tau_J + 1)} \right] \frac{A_1 K_T N}{K_D} \tag{5-6}$$

The coefficient of the s^2 term in the numerator is factored out, thereby putting the numerator in the "standard form" of the second order equation:

$$G(s) = \left[\frac{s^2 + 2\zeta' \omega_y s + \omega_y^2}{s^2 (s\tau_M + 1)(s\tau_f + 1)(s\tau_J + 1)} \right] K \tag{5-7}$$

where

$$K \triangleq \frac{A_1 K_T N D}{K_D} \tag{5-8}$$

$$\omega_y^2 \triangleq \frac{(k_\phi / \tau_2)}{D} \tag{5-9}$$

$$2\zeta' \omega_y \triangleq \frac{k_i + (\tau_1 + \tau_M) \frac{k_\phi}{\tau_2}}{D} \tag{5-10}$$

$$D \triangleq \left[k_m G_1 + k_i \tau_M + \tau_M \tau_1 \left(\frac{k_\phi}{\tau_2} \right) \right] \tag{5-11}$$

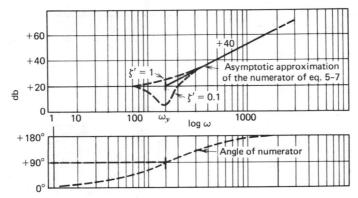

Figure 5-7 Bode plot of the numerator of eq. 5-7. The actual curve may differ substantially from the asymptotic approximation depending on the value of ζ'.

Some comments on the numerator of eq. 5-7 are important here. This numerator term, when observed in the pure frequency domain ($s = j\omega$), will have the general magnitude and angle plots of Fig. 5-7. Note that the "break frequency" occurs at $\omega = \omega_y$, and that the substantial "peaking" occurs for $\zeta' < 0.3$. (ζ' characterizes the "peaking" of the numerator term, and is not the system damping ratio ζ.)

Consider next the shape of the denominator term of eq. 5-7. A representative magnitude and angle plot (with $s = j\omega$) is shown in Fig. 5-8. It is desired to place ω_y.(the numerator break frequency) so as to have the phase margin (PM) to acceptable levels. Following the standard practice of Bode design, PM should be between 40 and 75°, depending on the amount of overshoot that can be accepted. When the two curves of Figs. 5-7 and 5-8 are superimposed, eq. 5-7 is truly represented and the general shape of the Bode plot for a properly designed system is as shown in Fig. 5-9.

Five significant frequencies can be identified:

ω_J = first break frequency caused by the inertia and damping ($= 1/\tau_J = K_D/J$)

ω_y = cut-in frequency of the numerator term

ω_c = crossover frequency of the Bode plot ($|GH(j\omega)| = 1 = 0$ dB)

ω_f = carrier filter frequency ($= 1/\tau_f$)

ω_M = monostable filter frequency ($= 1/\tau_M$)

Very often ω_c can be identified with the closed loop bandwidth of the system.

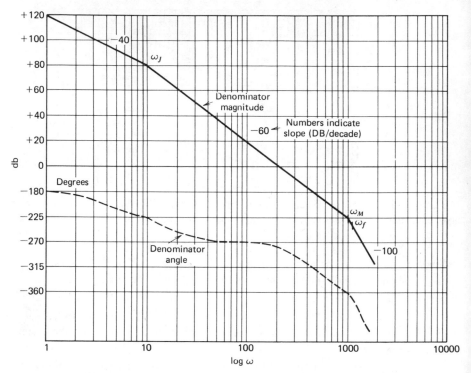

Figure 5-8 Bode plot of the denominator of eq. 5-7.

For a critically damped system ($\zeta = 1$) this is exactly true. For an underdamped case the bandwidth is somewhat larger than ω_c. Consequently, the use of ω_c as the closed loop bandwidth is a conservative estimate, and will in fact produce a somewhat faster response than predicted by equating ω_c and the bandwidth.

This implies that whenever bandwidth is specified, ω_c can automatically be set equal to bandwidth. However, some design problems may not directly give a bandwidth (or response time) specification. It is nonetheless necessary to know the crossover frequency of the open loop transfer function at this point in the design. To establish ω_c, use can be made of the inequality

$$\omega_c \leqslant \frac{\pi}{300} N \text{ rpm}_{\min} \tag{5-12}$$

(This equation is identical to eq. 4-3 except that the crossover frequency ω_c is used.) Any pair of the three variables ω_c, N, and $\text{rpm}_{\min}$ must now be specified.

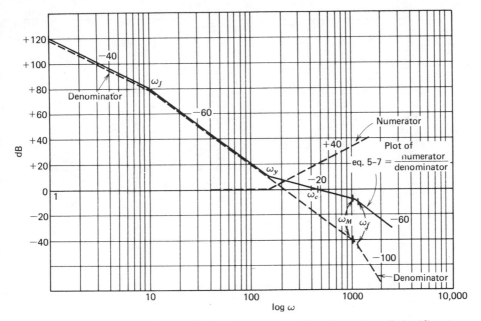

Figure 5-9 Bode plot of eq. 5-7 with appropriate locations for all significant break frequencies.

The third variable is then determined from eq. 5-12. N is selected to have certain standard values (typically between 100 and 5000 lines). Furthermore, the Bode plot must retain the general shape shown in Fig. 5-9. This implies that $\omega_J < \omega_y < \omega_c < \omega_f$ and ω_m. Stated another way, the bandwidth (ω_c), disc line density (N), and minimum shaft speed (rpm$_{min.}$) cannot be independently specified, and must satisfy eq. 5-12. The analysis proceeds on the assumption that these variables have been appropriately selected by the designer.

Consider next the damping term ζ' in the numerator of eq. 5-7. It must be chosen to avoid excessive peaking at ω_y. With excessive peaking an unexpected crossing of the 0 dB axis might occur. This is shown in Fig. 5-10. This problem is avoided by specifying $\zeta' = 1$, thus ensuring a well behaved lead network near ω_y.

It is often convenient to choose $\omega_M = \omega_f$. This permits maximum filtering of both the monostable signal and the total output signal (which contains carrier frequency components from both the D/A integrator and the op amp phase control circuit). There is no reason to favor one filter over the other, and ω_f is set equal to ω_M.

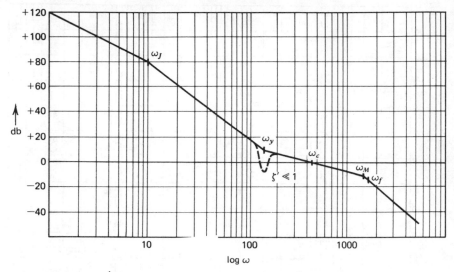

Figure 5-10 If ζ' is too small, excessive peaking of the numerator term may cause an undesired crossing of the 0 dB axis, thus causing instability.

It is good practice to place ω_y as much below ω_c as ω_M is above ω_c. If η is defined as the ratio of ω_M to ω_c, then

$$\omega_y = \frac{\omega_c}{\eta} \tag{5-13}$$

$$\omega_f = \omega_M = \eta\omega_c \tag{5-14}$$

The reason for the symmetrical placement of the two break frequencies is that the phase angle is a strongly nonlinear function of "distance" from ω_c. Near ω_c the phase angle changes rapidly. At some "distance" from ω_c the phase angle changes slowly. If ω_y and ω_M were unsymmetrically placed about ω_c, one of the break frequencies would be contributing most of the lag and the other break frequency would be relatively ineffective in producing the "lead" necessary to counteract the lag and achieve the required phase margin. Stated another way, if ω_M is placed very close to ω_c ($\omega_M = 3\,\omega_c$, for example), ω_y must be placed very far back from ω_c ($\omega_y = \omega_c/14$ for a 45° phase margin). In some cases it may be necessary to select ω_M as low as possible to obtain maximum filtering of the carrier frequency (ω_{tach}) at the lowest speed. A completely general program for PLL design allows independent selection of all the adjustable break frequencies. However, usually it is acceptable to assume symmetrical placement of ω_y and ω_M for the reasons outlined above. This will be done for the remainder of this chapter.

The equation for phase margin can be obtained from eq. 5-7, which is rewritten below for convenient reference:

$$G(s) = K\left[\frac{s^2 + 2\zeta'\omega_y s + \omega_y^2}{s^2(s\tau_M + 1)(s\tau_f + 1)(s\tau_J + 1)}\right] \tag{5-7}$$

In the pure frequency domain ($s = j\omega$), if $\zeta' = 1$ as discussed previously, eq. 5-7 can be rewritten as

$$G(j\omega) = K\left[\frac{(j\omega)^2 + 2\omega_y(j\omega) + \omega_y^2}{(j\omega)^2\left(j\frac{\omega}{\omega_M} + 1\right)\left(j\frac{\omega}{\omega_f} + 1\right)\left(j\frac{\omega}{\omega_J} + 1\right)}\right] \tag{5-15}$$

where $s = j\omega$

$$\omega_M = \frac{1}{\tau_M}$$

$$\omega_f = \frac{1}{\tau_f}$$

$$\omega_J = \frac{1}{\tau_J}$$

When real and imaginary terms are collected in the numerator, eq. 5-15 becomes

$$G(j\omega) = K\left[\frac{j2\omega_y\omega + (\omega_y^2 - \omega^2)}{(-\omega^2)\left(j\frac{\omega}{\omega_M} + 1\right)\left(j\frac{\omega}{\omega_f} + 1\right)\left(j\frac{\omega}{\omega_J} + 1\right)}\right] \tag{5-16}$$

On rewriting eq. 5-16 in terms of magnitude and angle, it follows that the polar form of the open loop transfer function is

$$X = |G(j\omega)| = \frac{K\sqrt{(2\omega\omega_y)^2 + (\omega_y^2 - \omega^2)^2}}{\omega^2\sqrt{(\omega/\omega_M)^2 + 1}\sqrt{(\omega/\omega_J)^2 + 1}\sqrt{(\omega/\omega_f)^2 + 1}} \tag{5-17}$$

$$\phi = \angle G(j\omega) = +\tan^{-1}\frac{2\omega\omega_y}{\omega_y^2 - \omega^2} - 180° - \tan^{-1}\frac{\omega}{\omega_M} - \tan^{-1}\frac{\omega}{\omega_J} - \tan^{-1}\frac{\omega}{\omega_f}$$

$$(5\text{-}18)$$

where X = magnitude of the open loop transfer function
ϕ = angle of the open loop transfer function

The general equation for the phase margin is

phase margin = PM = $\angle G(j\omega_c)$ + 180°   (ω_c = crossover frequency)   (5-19)

Using eq. 5-18 in eq. 5-19, we have the phase margin:

$$PM = +\tan^{-1}\frac{2\omega_c\omega_y}{\omega_y^2 - \omega_c^2} - \tan^{-1}\frac{\omega_c}{\omega_M} - \tan^{-1}\frac{\omega_c}{\omega_f} - \tan^{-1}\frac{\omega_c}{\omega_J} \quad (5\text{-}20)$$

When ω_f is set equal to ω_M (as per previous discussion) and ω_M and ω_y are placed symmetrically (eqs. 5-13 and 5-14 are used), the phase margin can be written as

$$PM = \tan^{-1}\frac{2\omega_c(\omega_c/\eta)}{(\omega_c/\eta)^2 - \omega_c^2} - 2\tan^{-1}\frac{\omega_c}{\eta\omega_c} - \tan^{-1}\frac{\omega_c}{\omega_J} \quad (5\text{-}21)$$

Upon simplification eq. 5-21 becomes

$$\boxed{\tan^{-1}\frac{2\eta}{1 - \eta^2} - 2\tan^{-1}\frac{1}{\eta} - \tan^{-1}\frac{\omega_c}{\omega_J} - PM = 0} \quad (5\text{-}22)$$

Equation 5-22 can be solved for η by using standard computer software for the "zeros of a function." By specifying the crossover frequency ω_c, ω_J (which implies specifying J and K_D) and the phase margin PM (which is a measure of stability), η is determined, and with it ω_y, ω_M, and ω_f.

Another important parameter K can be calculated from eq. 5-17. Because the magnitude of $G(j\omega)$ is unity at ω_c (by definition of ω_c), it may be written that

$$|G(j\omega_c)| = 1 = \frac{K\sqrt{(2\omega_y\omega_c)^2 + (\omega_y{}^2 - \omega_c{}^2)^2}}{\omega_c{}^2\sqrt{(\omega_c/\omega_J)^2 + 1}\sqrt{(\omega_c/\omega_M)^2 + 1}\sqrt{(\omega_c/\omega_f)^2 + 1}} \qquad (5\text{-}23)$$

Using eqs. 5-13 and 5-14 in eq. 5-23 results in

$$1 = \frac{K\sqrt{\left(2\left(\dfrac{\omega_c}{\eta}\right)\omega_c\right)^2 + \left(\left(\dfrac{\omega_c}{\eta}\right)^2 - \omega_c{}^2\right)^2}}{\omega_c{}^2\sqrt{(\omega_c/\omega_J)^2 + 1}\sqrt{(\omega_c/\omega_c\eta)^2 + 1}\sqrt{(\omega_c/\omega_c\eta)^2 + 1}} \qquad (5\text{-}24)$$

Solving for K:

$$K = \sqrt{\left(\frac{\omega_c}{\omega_J}\right)^2 + 1} \qquad (5\text{-}25)$$

Since ω_c and ω_J are known, K is now known from eq. 5-25. But another equation exists for K, namely eq. 5-8, which is shown here for convenient reference:

$$K = \frac{A_1 K_T ND}{K_D} \qquad (5\text{-}8)$$

Solving eq. 5-8 for D:

$$D = \frac{KK_D}{A_1 K_T N} \qquad (5\text{-}26)$$

This allows the calculation of D as a function of transconductance gain A_1 and K.

The complete solution for the remaining parameters depends on recognizing two facts about eqs. 5-9, 5-10, and 5-11. First, k_ϕ and τ_2 always appear together (as k_ϕ/τ_2). This implies that they can be represented by a single variable k_ρ such that

$$k_\rho = \frac{k_\phi}{\tau_2} \qquad (5\text{-}27)$$

k_ϕ and τ_2 are simply two controls for the same variable! Second, τ_1 should be set such that

$$\frac{1}{\tau_1} = \omega_y = \frac{\omega_c}{\eta} \tag{5-28}$$

This choice permits the fastest time constant for the phase restoration circuit (the op amp) without changing the basic nature of the Bode plot. If $1/\tau_1$ is adjusted beyond ω_y, a pole beyond ω_y results, and the Bode plot is "distorted."

With these considerations in mind, eqs. 5-13, 5-14, 5-27, and 5-28 are used in eqs. 5-9, 5-10, and 5-11 to obtain

$$\left(\frac{\omega_c}{\eta}\right)^2 = \frac{k_\rho}{D} \tag{5-29}$$

$$2\left(\frac{\omega_c}{\eta}\right) = \frac{k_i + (\eta + 1/\eta)(k_\rho/\omega_c)}{D} \tag{5-30}$$

$$D = k_m G_1 + \frac{k_i}{\eta\omega_c} + \left(\frac{1}{\eta\omega_c}\right)\frac{1}{(\omega_c/\eta)}k_\rho \tag{5-31}$$

On solving eqs. 5-29, 5-30, and 5-31 for k_ρ, k_i, and G_1, respectively, it is found that

$$k_\rho = \left(\frac{\omega_c}{\eta}\right)^2 D \tag{5-32}$$

$$k_i = 2\frac{\omega_c}{\eta}D - \frac{k_\rho}{\omega_c}\left(\eta + \frac{1}{\eta}\right) \tag{5-33}$$

$$G_1 = \frac{D - (k_i/\eta\omega_c) - (k_\rho/\omega_c^2)}{k_m} \tag{5-34}$$

These equations, along with their supporting relationships, completely specify the motor control PLL. Note that this analysis has used only the open loop transfer function to determine the operation of the closed loop. This is the beauty of the Bode method: it avoids excessive mathematical difficulty at the expense of a somewhat inexact stability criterion (the phase margin).

It is important to realize, however, that cases exist where a perfect Bode design yields a marginally stable or even unstable servo system! This is because the Bode plot examines the open loop transfer function at only one point, namely, at the crossover (0 dB) point. The Nyquist plot examines $G(j\omega)$ at all frequencies. It is useful to do a Nyquist plot to check the Bode design. The Nyquist plot is easily derived from eqs. 5-17 (magnitude) and 5-18 (angle). The computer or programmable calculator are invaluable aids in this otherwise formidable, laborious task.

CLOSED LOOP RESPONSE

The *closed loop* frequency response (CLR) is readily obtained from eqs. 5-17 and 5-18. Recall that

$$X = |G(j\omega)| = \text{magnitude of the open loop transfer function}$$

$$\phi = \angle G(j\omega) = \text{angle of the open loop transfer function}$$

The closed loop transfer function is given by

$$\text{closed loop transfer function} = F(j\omega) = \frac{G(j\omega)}{1 + G(j\omega)} \qquad (5\text{-}35)$$

and

$$|F(j\omega)| = \left| \frac{G(j\omega)}{1 + G(j\omega)} \right| = \frac{|G(j\omega)|}{|1 + G(j\omega)|} = \frac{X}{|1 + X_{\angle\phi}|} \qquad (5\text{-}36)$$

The denominator of eq. 5-36 can be portrayed in the complex plane as shown in Fig. 5-11. The sum $X_{\angle\phi} + 1$ can be written as

$$X_{\angle\phi} + 1 = (X\cos\phi + 1) + jX\sin\phi \qquad (5\text{-}37)$$

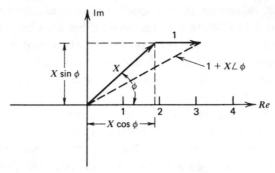

Figure 5-11 Phasor diagram of the denominator of eq. 5-36.

The magnitude of this complex number is

$$|X_{\underline{/\phi}} + 1| = \sqrt{(X\cos\phi + 1)^2 + (X\sin\phi)^2} \qquad (5\text{-}38)$$

Expanding, we have

$$|X_{\underline{/\phi}} + 1| = \sqrt{(X^2\cos^2\phi + 2X\cos\phi + 1) + X^2\sin^2\phi} \qquad (5\text{-}39)$$

Using the trigonometric identity $\sin^2\phi + \cos^2\phi = 1$ in eq. 5-39:

$$|X_{\underline{/\phi}} + 1| = \sqrt{X^2 + 2X\cos\phi + 1} \qquad (5\text{-}40)$$

It follows that the closed loop transfer function is

$$|F(j\omega)| = \frac{X}{\sqrt{X^2 + 2X\cos\phi + 1}} \qquad (5\text{-}41)$$

where X and ϕ are defined by eqs. 5-17 and 5-18 respectively.

The foregoing analysis establishes, (1) a means of designing the PLL motor control servo, (2) a means for checking the results, and (3) a means for determining the *closed* loop frequency response. The degree of peaking in $|F(j\omega)|$ (the closed loop response) is an indication of the overshoot and settling time of the resulting servo system. Obtaining numerical answers from this analysis is made possible by the digital computer. The labor involved *without* the computer is intimidating. For example, when the closed loop response is to be calculated, it is first necessary to obtain X and ϕ (from eqs. 5-17 and 5-18) and then substitute these values into eq. 5-41 to obtain $|F(j\omega)|$. This procedure is repeated at each new frequency where the response needs to be known. Going further back into the analysis, the equation for η (eq. 5-22) must be solved numerically or by trial and error. It is therefore a necessity to develop computer programs to obtain practical results in reasonable time. Flowcharts of the design process are presented in Charts 5-1 and 5-2. Chapter 6 makes use of these flowcharts to design programs for the PLL servo.

CHART 5-1   FLOWCHART FOR PLL DESIGN

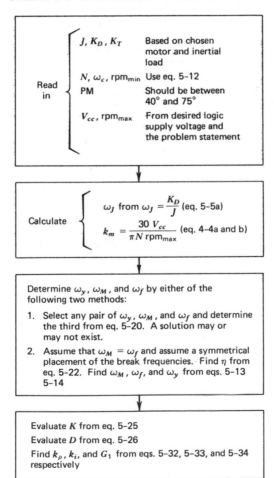

Read in
- J, K_D, K_T — Based on chosen motor and inertial load
- $N, \omega_c, \text{rpm}_{min}$ — Use eq. 5-12
- PM — Should be between $40°$ and $75°$
- V_{cc}, rpm_{max} — From desired logic supply voltage and the problem statement

Calculate
- ω_J from $\omega_J = \dfrac{K_D}{J}$ (eq. 5-5a)
- $k_m = \dfrac{30\, V_{cc}}{\pi N\, \text{rpm}_{max}}$ (eq. 4-4a and b)

Determine ω_y, ω_M, and ω_f by either of the following two methods:

1. Select any pair of ω_y, ω_M, and ω_f and determine the third from eq. 5-20. A solution may or may not exist.

2. Assume that $\omega_M = \omega_f$ and assume a symmetrical placement of the break frequencies. Find η from eq. 5-22. Find ω_M, ω_f, and ω_y from eqs. 5-13 5-14

Evaluate K from eq. 5-25
Evaluate D from eq. 5-26
Find k_ρ, k_i, and G_1 from eqs. 5-32, 5-33, and 5-34 respectively

**CHART 5-2   FLOWCHART FOR OBTAINING THE NYQUIST
AND THE CLOSED LOOP RESPONSE PLOTS**

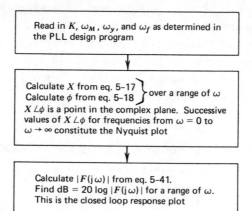

Read in K, ω_M, ω_y, and ω_f as determined in
the PLL design program

Calculate X from eq. 5–17 ⎱ over a range of ω
Calculate ϕ from eq. 5–18 ⎰
$X \angle \phi$ is a point in the complex plane. Successive
values of $X \angle \phi$ for frequencies from $\omega = 0$ to
$\omega \to \infty$ constitute the Nyquist plot

Calculate $|F(j\omega)|$ from eq. 5–41.
Find dB $= 20 \log |F(j\omega)|$ for a range of ω.
This is the closed loop response plot

6

Computer Aided Design

The equations and flowcharts developed in Chapter 5 are ready for use in selecting specific values for G_1, k_i, k_ρ, τ_M, and τ_f from some given performance specifications. However, the numerical complexity of the selection problem requires computer aid. The flowcharts make the results of Chapter 5 particularly suitable for computer aided design.

The machine for the computing tasks is the TI59 programmable calculator/ printer. It is selected for the following reasons.

1 The TI59 is adequate for this problem. The available "solid state software" contains a program that can be used to solve for η (see eq. 5-22).
2 The cost of the TI59, compared to any other machine presently available, is very low.
3 The TI59 is easy to use.
4 An understanding of the essence of the TI59 programs should make it simple to devise programs that will run on any computer, large or small.

Since the computational labor in velocity servo design is minimal, no program for such systems is given here. Velocity servo design is primarily a problem in resolving performance specifications.

Motor control PLL design, on the other hand, is a formidable task without the aid of the computer. The first program presented here is the "PLL DESIGN" program, which calculates the values of all parameters necessary for the motor control PLL. The second program, "ANGL, MAGN, CLR," permits the plotting of the open loop magnitude and angle (from which a Nyquist plot can be made) and the closed loop frequency response (which allows an estimate of the time domain performance). The program steps are listed in sequence to help the user understand the individual steps and/or to program any available computer. The appropriate flowcharts can be referred to as needed. Before proceeding to the program description, some notes on the format of this presentation may be helpful.

Each step is listed separately, together with explanatory notes and information about its location in the program. The numbers in parentheses adjacent to the step number indicate the location of the step in the TI59 program. For example:

Step 2 (000 to 011)

Indicates that step 2 occupies program locations 000 to 011. A complete program listing for both programs is given on pages 103 to 108. Furthermore, explanatory notes follow each program step as needed.

PROGRAM "PLL DESIGN"

Step 1 (No Program Space Used)

Store the data in memory (8 entries into memory) (see note 1)

J (oz-in-sec^2) into memory 10
K_D (oz-in./1000 pm) into memory 12 (note 2)
K_T (oz-in./A) into memory 14
PM (degrees) into memory 16
N (lines) into memory 18
V_{cc} (volts) into memory 20
HRPM(max$_{rpm}$) into memory 22
ω_c (rad/sec) into memory 24 (note 3)

Notes

1 A complete memory listing is given along with the program.
2 K_D is often specified with the units oz-in./1000 rpm. This is converted to oz-in./rad/sec in the program by multiplying by $3/100\,\pi$. Note that

$$1000 \text{ rpm} \times \frac{1 \text{ min}}{60 \text{ sec}} \times 2\pi \text{ rad/rev} = \frac{100\pi}{3} \text{ rad/sec}$$

3 Before entering these data, the relationship between ω_c, N, and rpm$_{min}$ must be chosen to satisfy eq. 5-12, which is shown below for convenient reference:

$$\omega_c \leqslant \frac{\pi}{300} N \text{ rpm}_{min} \tag{5-12}$$

The program assumes that eq. 5-12 has been satisfied and makes no verification thereof, except to calculate rpm$_{min}$ (sometimes designated LRPM) from ω_c and N.

Step 2 (000 to 011)

Print the heading "PLL DESIGN." This allows the user to immediately see that the correct program is being run and identifies the results for future examination.

Step 3 (012 to 027)

Calculate ω_J from

$$\omega_J = \frac{1}{\tau_J} = \frac{K_D}{J}$$

Note that K_D must have the units oz-in./rad/sec, while the data in memory have the units oz-in./1000 rpm. The unit conversion is done in the program. (See the notes to step 1.)

Step 4 (028 to 048 and Subroutine A', 349 to 397)

Calculate η. This calculation is done, using the TI59 Solid State Software Program 08, to solve eq. 5-22:

$$\tan^{-1}\frac{2\eta}{1-\eta^2} - 2\tan^{-1}\frac{1}{\eta} - \tan^{-1}\frac{\omega_c}{\omega_J} - PM = 0 \qquad (5\text{-}22)$$

Program 08 requires that an upper and lower limit be set for η. The determination of these limits begins by noting the η is always greater than unity. Making η less than unity reverses the relative positions of ω_f and ω_y about ω_c and is not applicable to this problem. The interval of search for a solution for η should therefore certainly have a lower limit greater than unity, thus avoiding both such reversal and the singularity point $\eta = 1$ (in the first term of eq. 5-22).

Furthermore, if the lower limit for η were to be chosen as low as $\eta = 1.1$, the greatest (best) phase margin that could be attained would be $10.9°$, which is too low a phase margin for any practical system. Consequently, η will always come out to be greater than 1.1 for any practical system. Hence if the search interval is chosen to start at $\eta = 1.1$, a solution will always exist (for phase margins better than $10.9°$) and the singularity point will always be avoided.

An upper limit of $\eta = 20$ implies that a maximum phase margin of $79°$ will be specified. Since the phase margin is usually specified between 35 and $75°$, $\eta = 20$ is beyond the limit for "normal" problems. The interval of search is therefore selected to be from $\eta = 1.1$ to $\eta = 20$.

The term $\tan^{-1}[2\eta/(1 - \eta^2)]$ will always yield a negative angle when evaluated for values of η greater than unity (as is presently the case). This is an incorrect result as far as this program is concerned! Referring to Fig. 5-7, note that the angle of the numerator function is always positive and is located between 0 and $180°$. The reason for the discrepancy lies in the multivalued nature of the arc tangent function. Consider Fig. 6-1. Angle $-x$ and angle $(180° - x)$

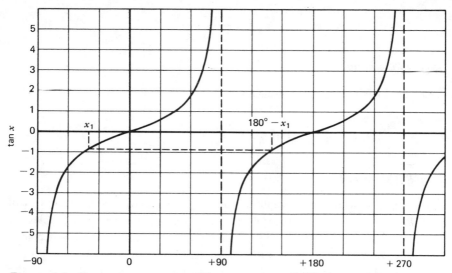

Figure 6-1 Tangent x versus x. The curve is periodic in x such that $\tan x = \tan(x + 180°)$. The calculator always gives results of the arc tangent operation in the region between -90 and $+90°$.

have the same tangent. In general, the calculator produces answers for the arc tan function that lie between $+90$ and $-90°$. This is only natural, since there is no convenient way to express the multivalued nature of the answer.

Because $2\eta/(1 - \eta^2)$ is evaluated only for $\eta \geqslant 1.1$, the term always has a negative value and the calculator always produces answers between 0 and $-90°$. To correct this, $180°$ is always added to the result to shift the range from between 0 and $-90°$ to between $+90$ and $+180°$. This is done in subroutine A', steps 392 through 395.

Step 5 (049 to 062)

Calculate ω_M and ω_y from eqs. 5-13 and 5-14:

$$\omega_y = \frac{\omega_c}{\eta} \tag{5-13}$$

$$\omega_M = \eta\omega_c \tag{5-14}$$

Step 6 (063 to 078)

Calculate k_m from eqs. 4-5 and 4-6:

$$k_m = \frac{30\,V_{cc}}{\pi N\,\mathrm{rpm}_{max}}$$

Step 7 (079 to 092)

Calculate LRPM (= rpm$_{min}$) using eq. 5-12:

$$\mathrm{rpm}_{min} = \frac{300\,\omega_c}{N\pi} \tag{5-12}$$

Step 8 (095 to 114)

User enters transductance amplifier gain, A_1. This allows indirect control of G_1, k_i, and k_ρ by direct control of A_1. In general, the former variables are inversely related to A_1. When A_1 is increased, G_1, k_i, and k_ρ tend to decrease. Inversely, decreasing A_1 causes the three variables to increase. By trying a few values for A_1, it is possible to bring G_1, k_i, and k_ρ to a "good" range. The program stops (at 115) to wait for user entry for A_1.

Step 9 (116 to 121 and Subroutine x^2, 240 to 255)

Calculate K and store results. Use eq. 5-25:

$$K = \sqrt{\left(\frac{\omega_c}{\omega_J}\right)^2 + 1} \tag{5-25}$$

Step 10 (122 to 125 and Subroutine $\sqrt{x}$, 256 to 282)

Calculate D and store results. Using eq. 5-26:

$$D = \frac{KK_D}{A_1 K_T N} \tag{5-26}$$

Step 11 (126 to 129 and Subroutine $1/x$, 283 to 293)

Calculate k_ρ. Use eq. 5-32:

$$k_\rho = \left(\frac{\omega_c}{\eta}\right)^2 D \tag{5-32}$$

Step 12 (130 to 133 and Subroutine RCL, 294 to 320)

Calculate k_i. Use eq. 5-33:

$$k_i = 2\frac{\omega_c}{\eta}D - \frac{k_\rho}{\omega_c}\left(\eta - \frac{1}{\eta}\right)$$ (5-33)

Step 13 (134 to 137 and Subroutine SUM, 321 to 348)

Calculate G_1. Use eq. 5-34:

$$G_1 = \frac{D - k_i/\eta\omega_c - k_\rho/\omega_c{}^2}{k_m}$$ (5-34)

Step 14 (138 to 170)

Print input variables and corresponding identification tags. The user must always know which input variables were used to obtain the output results.

Step 15 (171 to 208)

Print output variables and corresponding identification tags. Note that for both steps 14 and 15 the corresponding identification has been placed in memory adjacent to the variable. The program can thus be made to "loop" through the entire memory and simultaneously print data with appropriate labeling.

Step 16 (209 to 239)

Allow reentry of the program for another value of A_1 (i.e., repeat program from step 7 onward).

These steps constitute a written flowchart for the PLL design program. The actual T159 program is given on pages 103 to 105. The memory listing is given on page 105. Typical runs are shown in the design examples.

PROGRAM "ANGL, MAGN, CLR"

Now that the various design parameters have been established, it is desirable to make a Nyquist plot of the resulting servo system. This guarantees that no "surprises" exist in the design. The Bode plot, which was used for designing, examines the Nyquist plot at only one point, the point where the open loop transfer function is unity (i.e., the 0 dB crossover). This can, in extreme cases, be deceptive. For example, if the Nyquist plot has the shape shown in Fig. 6-2, substantial overshoot will occur even if the phase margin is "right." This is explained in greater detail in Appendix A.

It is also useful to determine the closed loop frequency response. The peaking of the closed loop frequency response plot is an indication of the time

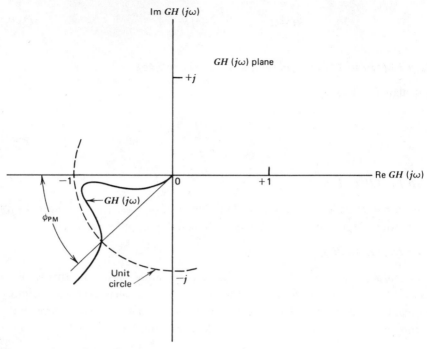

Figure 6-2 A "good" phase margin may yield a highly underdamped servo loop if the − 1 point is approached as shown.

domain damping of the system. To illustrate this relationship, Appendix E shows both time and frequency response curves for a second order system with a unit step input. The lower ζ becomes, the more pronounced will be the frequency domain peaking and the maximum time domain overshoot and decay time. Although the plots of Appendix E apply only to second order systems, they are useful for comparison. If a higher order system has the same frequency domain peaking as a second order system, the time domain transient response of the higher order system will be very similar, though not exactly identical, to the response of the second order system.

The program on pages 106 to 108 calculates the open loop transfer function magnitude and angle, and the closed loop response, from four input numbers, K, ω_M, ω_y, and ω_J. These parameters are determined in the design program previously described.

The quantities are calculated at frequencies in the ratio 1, 2, 5 beginning at $\omega = 10$ rad/sec. The 1, 2, 5 sequence is terminated after the first occurrence of a closed loop magnitude less than −7 dB. Stated another way, the program run ends the first time the closed loop frequency response falls below -7 dB.

However, the output quantities may be calculated (the program reentered) at any arbitrary frequency by simply entering the desired frequency into the display, and pressing "*A*". A description of the program is as follows.

Step 1 (No Program Space Used)

Enter parameters into memory. All the following parameters are available as output variables in the "PLL DESIGN" program.

$$K \text{ into memory } 02$$

$$\omega_M \text{ into memory } 38$$

$$\omega_y \text{ into memory } 36$$

$$\omega_J \text{ into memory } 34$$

Step 2 (000 to 017)

Print heading "ANGL, MAGN, CLR." This identifies the tape for future reference.

Step 3 (019 to 050)

Print input parameters with their identifying labels.

Step 4 (052 to 065)

Initialize frequencies at 10, 20, and 50 rad/sec.

Step 5 (073 to 080)

Print ω.

Step 6 (081 to 088 and Subroutine |x|, 132 to 196)

Calculate and print open loop magnitude. Use eq. 5-17:

$$|G(j\omega)| = X = \frac{K\sqrt{(2\omega\omega_y)^2 + (\omega_y{}^2 - \omega^2)^2}}{\omega^2\sqrt{((\omega/\omega_M)^2 + 1)((\omega/\omega_J)^2 + 1)((\omega/\omega_f)^2 + 1)}} \qquad (5\text{-}17)$$

Step 7 (089 to 098 and Subroutine cos, 197 to 271)

Calculate and print open loop angle. Use eq. 5-18:

$$\angle G(j\omega) = \phi = +\tan^{-1}\frac{2\omega\omega_y}{\omega_y{}^2 - \omega^2} - 180° - \tan^{-1}\frac{\omega}{\omega_M} - \tan^{-1}\frac{\omega}{\omega_J} - \tan^{-1}\frac{\omega}{\omega_f}$$

$$(5\text{-}18)$$

Special tests must be performed on the expression $\tan^{-1}[2\omega\omega_y/(\omega_y{}^2 - \omega^2)]$. First, if $\omega = \omega_y$, the function has a singularity that causes an error indication. Consequently, if $\omega = \omega_y$, the function must be considered as being equal to 90° and so added to the total angle. Step 212 tests for $\omega - \omega_y = 0$ and adds 90° if required (steps 239 to 241). Second, if $(\omega_y{}^2 - \omega^2) < 0$, 180° must be added to the angle as discussed earlier. This is done indirectly in the program by first testing for $(\omega_y{}^2 - \omega^2) < 0$ (steps 226 to 227) and *not subtracting* 180° if the condition is true.

Step 8 (099 to 106 and Subroutine lnx, 272 to 302)

Calculate and print CLR. Use Eq. 5-41:

$$F(j\omega) = \frac{X}{\sqrt{X^2 + 2X\cos\phi + 1}} \tag{5-41}$$

Step 9

Reenter program at "label tan" (066) to evaluate any frequency ω.

Typical program results are shown in the design example. These two programs allow the design, and parameter verification, of the motor control PLL of the form shown in Fig. 5-6. Some design examples are presented to illustrate the use of the programs.

PLL DESIGN

Print "PLL DESIGN"	000	69	OP		058	43	RCL		116	42	STO
	001	00	00		059	30	30		117	26	26
	002	43	RCL		060	95	=	K	118	71	SBR
	003	48	.48		061	42	STO		119	33	X²
	004	69	OP		062	36	36		120	42	STO
	005	01	.01		063	43	RCL		121	46	46
	006	43	RCL		064	20	20		122	71	SBR
	007	49	.49		065	65	×	D	123	34	ΓX
	008	69	OP		066	03	3		124	42	STO
	009	02	02		067	00	0		125	03	03
	010	69	OP		068	55	÷		126	71	SBR
	011	05	05		069	43	RCL		127	35	1/X
	012	03	3	km	070	18	18	kp	128	42	STO
	013	55	÷		071	55	÷		129	40	40
	014	01	1		072	43	RCL		130	71	SBR
	015	00	0		073	22	22		131	43	RCL
	016	00	0		074	55	÷	ki	132	42	STO
	017	55	÷		075	89	π		133	42	42
	018	89	π		076	95	=		134	71	SBR
	019	65	×		077	42	STO		135	44	SUM
	020	43	RCL		078	28	28	G1	136	42	STO
ωj	021	12	12		079	03	3		137	44	44
	022	55	÷		080	00	0		138	69	OP
	023	43	RCL		081	00	0		139	00	00
	024	10	10		082	55	÷		140	43	RCL
	025	95	=		083	89	π		141	50	50
	026	42	STO		084	65	×	Print "INPUT"	142	69	OP
	027	34	34		085	43	RCL		143	01	01
	028	01	1		086	24	24		144	69	OP
	029	93	.	LRPM	087	55	÷		145	05	05
	030	01	1		088	43	RCL		146	02	2
	031	36	PGM		089	18	18		147	07	7
	032	08	08		090	95	=		148	42	STO
	033	11	A		091	42	STO		149	00	00
	034	02	2		092	32	32		150	09	9
	035	00	0		093	76	LBL		151	32	X:T
	036	36	PGM		094	10	E'		152	76	LBL
	037	08	08		095	69	OP	Print loop	153	38	SIN
	038	12	B		096	00	00		154	73	RC*
	039	01	1		097	43	RCL		155	00	00
	040	00	0		098	58	58		156	69	OP
Find η	041	36	PGM		099	69	OP		157	04	04
	042	08	08		100	01	01		158	69	OP
	043	13	C		101	43	RCL		159	30	30
	044	36	PGM	Print "ENTER A1, PRESS RS"	102	53	53		160	73	RC*
	045	08	08		103	69	OP		161	00	00
	046	15	E		104	02	02		162	69	OP
	047	42	STO		105	43	RCL		163	06	06
	048	30	30		106	56	56		164	69	OP
	049	65	×		107	69	OP		165	30	30
	050	43	RCL		108	03	03		166	43	RCL
	051	24	24		109	43	RCL		167	00	00
	052	95	=		110	59	59		168	22	INV
ωM	053	42	STO		111	69	OP		169	67	EQ
Wy	054	38	38		112	04	04		170	38	SIN
	055	43	RCL		113	69	OP				
	056	24	24		114	05	05				
	057	55	÷	Wait	115	91	R/S				

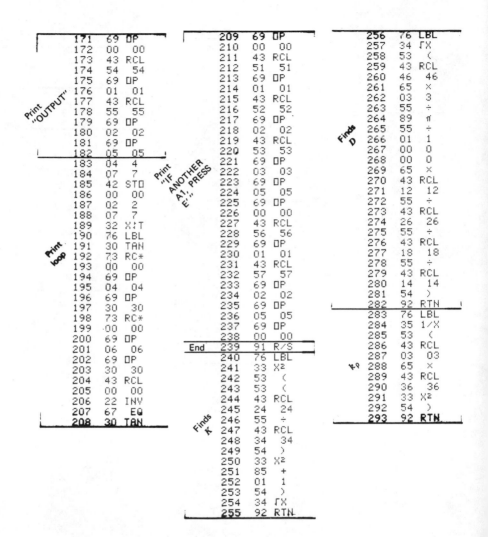

171	69	OP		209	69	OP		256	76	LBL
172	00	00		210	00	00		257	34	√X
173	43	RCL		211	43	RCL		258	53	(
174	54	54		212	51	51		259	43	RCL
175	69	OP		213	69	OP		260	46	46
176	01	01		214	01	01		261	65	×
177	43	RCL		215	43	RCL		262	03	3
178	55	55		216	52	52		263	55	÷
179	69	OP		217	69	OP		264	89	π
180	02	02		218	02	02		265	55	÷
181	69	OP		219	43	RCL		266	01	1
182	05	05		220	53	53		267	00	0
183	04	4		221	69	OP		268	00	0
184	07	7		222	03	03		269	65	×
185	42	STO		223	69	OP		270	43	RCL
186	00	00		224	05	05		271	12	12
187	02	2		225	69	OP		272	55	÷
188	07	7		226	00	00		273	43	RCL
189	32	X:T		227	43	RCL		274	26	26
190	76	LBL		228	56	56		275	55	÷
191	30	TAN		229	69	OP		276	43	RCL
192	73	RC*		230	01	01		277	18	18
193	00	00		231	43	RCL		278	55	÷
194	69	OP		232	57	57		279	43	RCL
195	04	04		233	69	OP		280	14	14
196	69	OP		234	02	02		281	54	)
197	30	30		235	69	OP		282	92	RTN
198	73	RC*		236	05	05		283	76	LBL
199	00	00		237	69	OP		284	35	1/X
200	69	OP		238	00	00		285	53	(
201	06	06		239	91	R/S		286	43	RCL
202	69	OP		240	76	LBL		287	03	03
203	30	30		241	33	X²		288	65	×
204	43	RCL		242	53	(		289	43	RCL
205	00	00		243	53	(		290	36	36
206	22	INV		244	43	RCL		291	33	X²
207	67	EQ		245	24	24		292	54	)
208	30	TAN		246	55	÷		293	92	RTN
				247	43	RCL				
				248	34	34				
				249	54	)				
				250	33	X²				
				251	85	+				
				252	01	1				
				253	54	)				
				254	34	√X				
				255	92	RTN				

Print "OUTPUT"

Print loop

Print "IF ANOTHER A1, PRESS E'"

End 239

Finds K

Finds D

1/X

294	76	LBL		349	76	LBL		27.		00
295	43	RCL		350	16	A'		11.1		01
296	53	(		351	42	STO		20.		02
297	02	2		352	30	30		.0001546667		03
298	65	×		353	53	(		5.02578125		04
299	43	RCL		354	53	(		5.035546875		05
300	36	36		355	43	RCL		5.030664063		06
301	65	×		356	24	24		-124.0898297		07
302	43	RCL		357	55	÷		0.01		08
303	03	03		358	43	RCL		0.		09
304	75	-		359	34	34		0.522	J	10
305	53	(		360	54	)		25000000.	"J"	11
306	43	RCL		361	22	INV		4.7	K_D	12
307	30	30		362	30	TAN		26160000.	"KD"	13
308	35	1/X		363	94	+/-		27.	K_T	14
309	85	+		364	85	+		26370000.	"KT"	15
310	43	RCL		365	53	(		45.	PM	16
311	30	30		366	02	2		33300000.	"PM"	17
312	54	)		367	65	×		5000.	N	18
313	65	×		368	43	RCL		31000000.	"N"	19
314	43	RCL		369	30	30		15.	V_{CC}	20
315	40	40		370	55	÷		42151500.	"VCC"	21
316	55	÷		371	53	(		3000.	HRPM	22
317	43	RCL		372	01	1		23353330.	"HRPM"	23
318	24	24		373	75	-		1000.	ω_C	24
319	54	)		374	43	RCL		43150000.	"WC"	25
320	92	RTN		375	30	30		25.	A_1	26
321	76	LBL		376	33	X²		13020000.	"A1"	27
322	44	SUM		377	54	)		.0000095493	k_m	28
323	53	(		378	54	)		26300000.	"KM"	29
324	53	(		379	22	INV		5.030664063	η	30
325	43	RCL		380	30	TAN		17371300.	"ETA"	31
326	42	42		381	75	-		19.09859317	LRPM	32
327	65	×		382	02	2		27353330.	"LRPM"	33
328	43	RCL		383	65	×		.0859802566	ω_J	34
329	38	38		384	43	RCL		43250000.	"WJ"	35
330	35	1/X		385	30	30		198.7809139	ω_y	36
331	85	+		386	35	1/X		43450000.	"WY"	37
332	43	RCL		387	22	INV		5030.664063	ω_M	38
333	24	24		388	30	TAN		43300000.	"WM"	39
334	35	1/X		389	75	-		6.111475759	k_ρ	40
335	33	X²		390	43	RCL		26352332.	"KRHO"	41
336	65	×		391	16	16		.0295299367	k_i	42
337	43	RCL		392	85	+		26240000.	"KI"	43
338	40	40		393	01	1		14.94195959	G_1	44
339	75	-		394	08	8		22020000.	"G1"	45
340	43	RCL		395	00	0		11630.5771	K	46
341	03	03		396	54	)		26000000.	"K"	47
342	54	)		397	92	RTN		3327270016.	PLL-D	48
343	55	÷						1736242231.	ESIGN	49
344	43	RCL						2431334137.	INPUT	50
345	28	28						2421001331.	IF-AN	51
346	94	+/-						3237231735.	OTHER	52
347	54	)						13025700.	-A1,-	53
348	92	RTN						3241373341.	OUTPU	54
								3700000000.	T····	55
								3335173636.	PRESS	56
								17650000.	-E'--	57
								1731371735.	ENTER	58
								35360000.	-RS··	59

k_i (label at step 305/306)

G_1 (label at step 335/336)

For η (label at column 2)

Program
"ANGL, MAGN, CLR"

Print heading

```
000   69  OP
001   00  00
002   43  RCL
003   59  59
004   69  OP
005   01  01
006   43  RCL
007   58  58
008   69  OP
009   02  02
010   43  RCL
011   57  57
012   69  OP
013   03  03
014   69  OP
015   05  05
016   69  OP
017   00  00
```

K

```
018   98  ADV
019   43  RCL
020   55  55
021   69  OP
022   04  04
023   43  RCL
024   02  02
025   69  OP
026   06  06
```

ωM

```
027   43  RCL
028   39  39
029   69  OP
030   04  04
031   43  RCL
032   38  38
033   69  OP
034   06  06
```

ωJ

```
035   43  RCL
036   37  37
037   69  OP
038   04  04
039   43  RCL
040   36  36
041   69  OP
042   06  06
```

ωY

```
043   43  RCL
044   35  35
045   69  OP
046   04  04
047   43  RCL
048   34  34
049   69  OP
050   06  06
```

Initialize

```
051   98  ADV
052   01  1
053   00  0
054   42  STO
055   09  09
056   02  2
057   00  0
058   42  STO
059   08  08
060   05  5
061   00  0
062   42  STO
063   07  07
064   07  7
065   32  X:T
066   76  LBL
067   30  TAN
068   09  9
069   42  STO
070   06  06
071   76  LBL
072   38  SIN
```

Print ω

```
073   43  RCL
074   56  56
075   69  OP
076   04  04
077   73  RC*
078   06  06
079   69  OP
080   06  06
```

Calculate and print MAGN

```
081   43  RCL
082   58  58
083   69  OP
084   04  04
085   71  SBR
086   50  IxI
087   69  OP
088   06  06
```

```
Calculate and print ±
089  43 RCL
090  59  59
091  69 OP
092  04  04
093  71 SBR
094  39 COS
095  43 RCL
096  04  04
097  69 OP
098  06  06

Calculate and print CLR
099  43 RCL
100  57  57
101  69 OP
102  04  04
103  71 SBR
104  23 LNX
105  69 OP
106  06  06

Test if reentered
107  87 IFF
108  01  01
109  35 1/X

Test and reinitialize and return
110  94 +/-
111  77 GE
112  35 1/X
113  69 OP
114  36  36
115  43 RCL
116  06  06
117  77 GE
118  38 SIN
119  01  1
120  00  0
121  49 PRD
122  09  09
123  49 PRD
124  08  08
125  49 PRD
126  07  07
127  61 GTO
128  30 TAN

End
129  76 LBL
130  35 1/X
131  91 R/S

132  76 LBL
133  50 IxI
134  53  (
135  53  (
136  53  (
137  02  2
138  65  x
139  73 RC*
140  06  06
141  65  x
142  43 RCL
143  36  36
144  54  )
145  33 X²

Calculate MAGN
146  85  +
147  53  (
148  43 RCL
149  36  36
150  33 X²
151  75  -
152  73 RC*
153  06  06
154  33 X²
155  54  )
156  33 X²
157  54  )
158  34 ΓX
159  55  ÷
160  73 RC*
161  06  06
162  33 X²
163  55  ÷
164  53  (
165  53  (
166  73 RC*
167  06  06
168  55  ÷
169  43 RCL
170  38  38
171  54  )
172  33 X²
173  85  +
174  01  1
175  54  )
176  55  ÷
177  53  (
178  53  (
179  73 RC*
180  06  06
181  55  ÷
182  43 RCL
183  34  34
184  54  )
185  33 X²
186  85  +
187  01  1
188  54  )
189  34 ΓX
190  65  x
191  43 RCL
192  02  02
193  54  )
194  42 STO
195  05  05
196  92 RTN

197  76 LBL
198  39 COS
199  00  0
200  32 X!T                         Test if denominator is zero
201  53  (
202  53  (
203  53  (
204  43 RCL
205  36  36
206  33 X²
207  75  -
208  73 RC*
209  06  06
210  33 X²
211  54  )
212  67 EQ
213  49 PRD
214  35 1/X
215  65  x
216  02  2
217  65  x
218  73 RC*                         tan⁻¹ < 0
219  06  06
220  65  x
221  43 RCL                         Test if tan⁻¹ < 0
222  36  36
223  54  )
224  22 INV
225  30 TAN
226  22 INV
227  77 GE
228  43 RCL
229  75  -
230  01  1
231  08  8
232  00  0
233  76 LBL        Calculate ± G(jω)
234  43 RCL
235  61 GTO
236  48 EXC
237  76 LBL
238  49 PRD
239  75  -
240  09  9
241  00  0
242  76 LBL
243  48 EXC
244  75  -
245  02  2
246  65  x
247  53  (
248  73 RC*
249  06  06
250  55  ÷
251  43 RCL
252  38  38
253  54  )
254  22 INV
```

107

#	code	op		register	addr
				0.	00
				0.	01
				490.K	02
				0.	03
255	30	TAN		-159.5174656	04
256	75	-		.2412635629	05
257	53	(		9.	06
258	73	RC*		5000.	07
259	06	06		2000.	08
260	55	÷		3000.	09
261	43	RCL		0.	10
262	34	34		0.	11
263	54	)		0.	12
264	22	INV		0.	13
265	30	TAN		0.	14
266	54	)		0.	15
267	42	STO		0.	16
268	04	04		0.	17
269	07	7		0.	18
270	32	X:T		0.	19
271	92	RTN		0.	20
272	76	LBL		0.	21
273	23	LNX		0.	22
274	53	(		0.	23
275	53	(		0.	24
276	43	RCL		0.	25
277	05	05		0.	26
278	55	÷		0.	27
279	53	(		0.	28
280	43	RCL		0.	29
281	05	05		0.	30
282	33	X²		0.	31
283	85	+		0.	32
284	02	2		0.	33
285	65	×		$2.\,\omega_J$	34
286	43	RCL		43250000. "WJ"	35
287	05	05		$200.\,\omega_y$	36
288	65	×		43450000. "WY"	37
289	43	RCL		$5000.\,\omega_M$	38
290	04	04		43300000. "WM"	39
291	39	COS		0.	40
292	85	+		0.	41
293	01	1		0.	42
294	54	)		0.	43
295	34	ΓX		0.	44
296	54	)		0.	45
297	28	LOG		0.	46
298	65	×		0.	47
299	02	2		0.	48
300	00	0		0.	49
301	54	)		0.	50
302	92	RTN		0.	51
303	76	LBL		0.	52
304	11	A		0.	53
305	42	STO		0.	54
306	09	09		26000000.K	55
307	86	STF		43000000.ω	56
308	01	01		5715273500. , CLR-	57
309	61	GTO		5730132231. , MAGN	58
310	30	TAN		13312227. —ANGL	59

Calculate CLR (steps near 285)

Reenter with random ω (steps near 306)

EXAMPLES

Design Example 1

A PMI U12M4H with 5000 line optics is to be phaselocked at speeds from 30 to 3000 rpm. The available CMOS power supply is +15 V. The inertial load consists only of the motor and the disc.

From the problem statement:

$$N = 5000$$

$$LRPM = 30$$

$$HRPM = 3000$$

$$V_{cc} = 15 \text{ V DC}$$

From the data sheet for a PMI U12M4H:

$$K_T = 27 \text{ oz-in/A}$$

$$K_D = 4.7 \text{ oz-in./1000 rpm}$$

$$J_{motor} = 0.02 \text{ oz-in-sec}^2$$

The inertia of the optics disc is known to be 0.002 oz-in-sec^2:

$$J_{optics} = 0.002 \text{ oz-in-sec}^2$$

The total inertia is the sum of the motor and disc inertias:

$$J = J_{motor} + J_{disc} = 0.022 \text{ oz-in-sec}^2$$

It is necessary to select an appropriate crossover frequency ω_c. This is done from eq. 5-12:

$$\omega_c \leqslant \frac{\pi}{300} N \text{LRPM} = \frac{\pi}{300} \times 5000 \times 30 = 1570 \text{ rad/sec} \qquad (5\text{-}12)$$

ω_c is selected to be 1000 rad/sec, thereby allowing a somewhat lower minimum rpm than required.

The following variables are entered into memory as per step 1 of the "PLL DESIGN" program:

$$0.022 \rightarrow \text{M10} \; (J, \text{oz-in-sec}^2)$$

$$4.7 \rightarrow \text{M12} \; (K_D, \text{oz-in.}/1000 \text{ rpm})$$

$$27 \rightarrow \text{M14} \; (K_T, \text{oz-in.}/\text{A})$$

$$45° \rightarrow \text{M16} \; (\text{phase margin, degrees})$$

$$5000 \rightarrow \text{M18} \; (\text{disc density})$$

$$15 \rightarrow \text{M20} \; (V_{cc}, \text{volts})$$

$$3000 \rightarrow \text{M22} \; (\text{HRPM})$$

$$1000 \rightarrow \text{M24} \; (\omega_c, \text{rad/sec})$$

The program is run by pressing "RST" and then "RS." A minute or two elapses, and the following printout appears:

```
PLL DESIGN
ENTER A1, PRESS RS
```

At this point, the program is "asking" for a value for A_1, the transconductance amplifier gain. The overall servo gain is not affected by the choice of A_1 because the servo gain is independent of A_1 (see eq. 5-25). A_1 merely determines the gain distribution, or which part of the system provides the gain. (Refer to the discussion of step 8 above.) As a first attempt, let $A_1 = 100$. (This gain level is commonly available on servo amplifiers.)

A value of 100 is entered into the display, and "RS" is pressed. The resulting printout is:

```
                             OUTPUT
    PLL DESIGN               490.1785972    K
    ENTER A1, PRESS RS       .1573336118    G1
    INPUT                    .0003122522    KI
            100.    A1       .0648960091    KRHO
           1000.    WC       5011.132813    WM
           3000.    HRPM     199.5556768    WY
             15.    VCC      2.040076998    WJ
           5000.    N        19.09859317    LRPM
             45.    PM       5.011132813    ETA
             27.    KT       .0000095493    KM
            4.7     KD       IF ANOTHER A1,
           0.022    J        PRESS E
```

It is observed that the gains G_1, k_i, and k_ρ are all somewhat low. This can be seen by calculating k_i in volts per count. In the printout, k_i is given as 0.31 mV/rad. Multiplying this value by the factor 2π converts to volts per count.

(There are 2π rad/count.) It follows that 0.31 mV/rad $\times$ 2π rad/count = 2 mV/count. This shows that the D/A converter output swing is only 2 mV in the steady state (see Fig. 3-11). Such a low signal level may be subject to noise contamination. As discussed above, A_1 will be decreased, and it is expected that G_1, k_i, and k_ρ will all increase. This should provide a more favorable gain distribution.

To obtain a printout for a new value of A_1, it is necessary to press "2nd," "E" ($= E'$), wait for the cue, enter the new value for A_1 into the display, and press "RS." It is not necessary to press "RST" when A_1 is changed. If "RST" were pressed, the computer would unnecessarily recalculate η and thereby take longer to produce results. Using $A_1 = 25$, it is found that:

```
ENTER A1, PRESS RS      OUTPUT
INPUT                   490.1785972    K
        25.    A1       .6293344472    G1
      1000.    WC       .0012490086    KI
      3000.    HRPM     .2595840363    KRHO
        15.    VCC      5011.132813    WM
      5000.    N        199.5556768    WY
        45.    PM       2.040076998    WJ
        27.    KT       19.09859317    LRPM
        4.7    KD       5.011132813    ETA
       0.022   J        .0000095493    KM
                        IF ANOTHER A1,
                        PRESS E'
```

The gain has been redistributed to increase G_1, k_i, and k_ρ. Note that K, the overall servo gain, is unchanged. Note also that $k_i = 1.2$ mV/rad $\times$ 2π rad/count $\cong 7.5$ mV/count.

The following observations can be made.

1 LRPM ($= \text{rpm}_{min}$) $= 19$ rpm. This meets specification.
2 A Bode plot is drawn in Fig. 6-3 from the data on ω_c, ω_M, ω_y, and ω_J. The plot is constructed by projecting at -20 dB/decade from ω_c ($\simeq 1000$). When the projection reaches ω_y ($\simeq 200$) and ω_M ($\simeq 5000$), a new slope of -60 dB/decade is used.

To check these results, the program "ANGL, MAGN, CLR" is loaded into the computer. Rounded data are entered as

$$490 \rightarrow \text{M02} \ (K, \text{dimensionless})$$
$$5000 \rightarrow \text{M38} \ (\omega_M, \text{rad/sec})$$
$$200 \rightarrow \text{M36} \ (\omega_y, \text{rad/sec})$$
$$2 \rightarrow \text{M34} \ (\omega_J, \text{rad/sec})$$

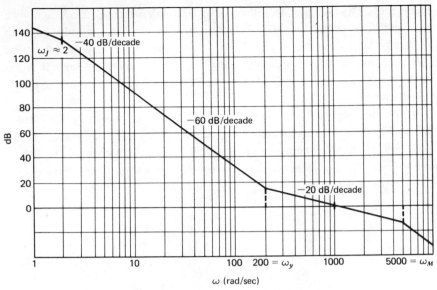

Figure 6-3 Bode plot for Example 1.

By pressing "RS," the following results are obtained:

```
ANGL, MAGN, CLR

          490.       K
         5000.       WM            200.         W
          200.       WY      9.783855867       MAGN
            2.       WJ     -184.0082814        ANGL
                             .9338036145        CLR
           10.       W            500.          W
    38534.70525      MAGN     2.2510711        MAGN
   -253.1944399      ANGL  -144.7948234        ANGL
     .0000651675     CLR     3.266253504        CLR
           20.       W           1000.          W                750.         W
     4924.360265     MAGN     0.97999804       MAGN          1.3687826      MAGN
   -253.3265844      ANGL  -135.1251385        ANGL       -136.7715772      ANGL
     .0005059306     CLR     2.254752364        CLR         3.287311971      CLR
           50.       W           2000.          W               1500.        W
     332.9004694     MAGN     .4266377177      MAGN          .6100436304     MAGN
   -240.7827804      ANGL  -144.9667095        ANGL       -138.5113809       ANGL
     .0127152672     CLR    -4.241359585        CLR         -.9033455062     CLR
          100.       W           5000.          W               3000.        W
     48.97061469     MAGN     .0981567921      MAGN          .2412635629     MAGN
   -218.0156605      ANGL  -184.5583018        ANGL       -159.5174656       ANGL
     .1401659285     CLR   -19.26752807        CLR        -10.17623056       CLR
```

The additional points for ω = 750, 1500, and 3000 were obtained by entering the frequency and pressing "A." The Nyquist and closed loop response plots are given in Figs. 6-4 and 6-5, respectively. The Nyquist plot shows a well

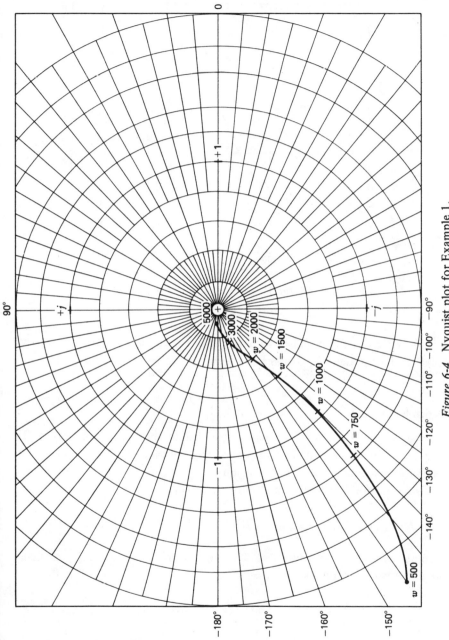

Figure 6-4 Nyquist plot for Example 1.

113

__114   Computer Aided Design__

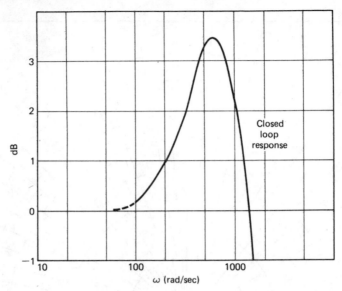

Figure 6-5 Closed loop response curve for Example 1.

behaved system with no hidden liabilities. The closed loop response plot shows only 3.5 dB of peaking. This indicates a well damped system.

Design Example 2

Assume that a ½ oz-in-sec^2 flywheel is added to the shaft of the U12M4H, with all other design requirements remaining the same as in Example 1. Determine appropriate values for G_1, k_i, and k_ρ.

From the problem statement, the total inertia is

$$J = J_{motor} + J_{disc} + J_{load} = 0.02 + 0.002 + 0.5 = 0.522 \text{ oz-in-sec}^2$$

All other parameters are identical. This one change is entered by simply changing the contents of memory 10:

$$0.522 \rightarrow M10 \ (J, \text{ oz-in-sec}^2)$$

No other memory is changed. Using $A = 100$ and running the program from its beginning by pressing "RST" and then "RS";

```
PLL DESIGN                      11630.5771      K
ENTER A1, PRESS RS              3.735489897     G1
INPUT                            .0073824842    KI
            100.    A1          1.52786894      KRHO
           1000.    WC          5030.664063     WM
           3000.    HRPM        198.7809139     WY
             15.    VCC          .0859802566    WJ
           5000.    N           19.09859317     LRPM
             45.    PM          5.030664063     ETA
             27.    KT           .0000095493    KM
              4.7   KD          IF ANOTHER A1,
              0.522 J           PRESS E'
```

These results are satisfactory in that "reasonable" numbers for G_1, k_i, and k_ρ are required. A gain G_1 of 3.7 is easy to attain; k_i of 7.38 mV/rad = 46 mV/count is reasonable; and $k_\rho = 1.5$ is also reasonable. The only "fault" with the system is that the gains must all be changed when the 0.5 oz-in-sec^2 inertia is used. An attempt to use the same A_1 (= 25) as in the low inertia case leads to the following printout (press "E," enter "25," press "RS"):

```
ENTER A1, PRESS RS      OUTPUT
INPUT                     11630.5771      K
             25.    A1    14.94195959     G1
           1000.    WC     .0295299367    KI
           3000.    HRPM  6.111475759     KRHO
             15.    VCC   5030.664063     WM
           5000.    N     198.7809139     WY
             45.    PM     .0859802566    WJ
             27.    KT    19.09859317     LRPM
              4.7   KD    5.030664063     ETA
              0.522 J      .0000095493    KM
                        IF ANOTHER A1,
                        PRESS E'
```

Comparing the values for G_1, k_i, and k_ρ in the low inertia case (Example 1) and the high inertia case (Example 2) shows that these values differ substantially:

	$A_1 = 25$	
	Example 1 $J = 0.022$ oz-in-sec^2	Example 2 $J = 0.522$ oz-in-sec^2
G_1	0.629	14.9
k_i	1.25 mV/rad	29.5 mV/rad
k_ρ	0.26	6.11

When the values of G_1, k_i, and k_ρ are adjusted to suit Example 1, the addition of an inertial load will probably cause instability unless all three parameters are readjusted as shown for Example 2. Under some circumstances, however, it is possible to design a PLL that is stable with both inertial loads (and all intermediate inertial loads) of Design Examples 1 and 2. Such a procedure is described in Appendix C.

7

Measurement and Practice

The theoretical development presented in the preceding chapters provides an excellent starting point for obtaining a working PLL system. However, it is necessary to establish a means of measuring performance and determining the quality of the lock. A number of measurement techniques will be described. Furthermore, in some circumstances the block diagram model of the PLL assumed throughout this book may not represent accurately the "real' life" system. Phenomena such as torsional resonance, noise, and disc jitter may modify or distort the assumed system model. Though correctly designed, systems may not work as planned. Some common problems and their solutions are detailed here to aid the user in translating a good design into functional hardware.

OSCILLOSCOPE DISPLAY TECHNIQUES

An excellent way of displaying phaselock system performance is with a dual trace oscilloscope (Fig. 7-1). Channel 1 is used to display the reference frequency (f_{ref}). Channel 2 is used to display the tachometer output frequency (f_{tach}). The scope is synchronized on channel 1, and a stable waveform can always be observed on channel 1. When the PLL is out of lock, the channel 2 trace is a blurr because it is not of the same frequency as the channel 1 signal. When phase or frequency lock is achieved, the tachometer signal (f_{tach}) "snaps" into alignment with the reference signal. Channel 2 is no longer a blurr; the signal has the same frequency as the reference, and has a definite phase relation to the reference frequency. (The scope should be in the "chop" mode for this measurement.)

This simple scheme is a reliable way of detecting and displaying phaselock. Furthermore, the channel 2 waveform can be used to provide an indication of

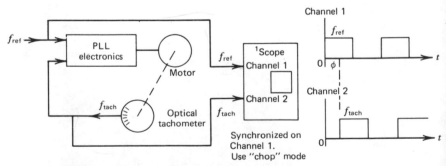

Figure 7-1 Test setup block diagram for display of PLL waveforms.

the "quality" of the lock. If a mechanical disturbance or retarding torque is applied to the motor shaft, the relative phase of the waveforms changes, even if only for a short time. The amount of the change represents the "stiffness" of the system. The recovery time, or how long it takes to regain the initial phase relationship, is a qualitative indication of the relative stability and damping of the servo. For example, if the channel 2 waveform is observed to oscillate about the initial phase (with the channel 1 display), the servo is underdamped.

As explained in Chapter 2, some FM is always present on the tachometer signal. The FM is due to residual misalignments of the disc, pattern imperfections, and actual speed changes of the shaft. The FM manifests itself as "jitter," or blurred leading and trailing edges of the tachometer waveform when displayed as described above. Such jitter is not apparent when the tachometer signal alone is displayed. A sketch of a typical scope waveform is shown in Fig. 7-2.

A quantitative measurement "percentage jitter" may be defined as

$$\% \text{ jitter} = \frac{\tau}{T} \times 100\% = \frac{\pm(\tau/2)}{T} \times 100\%$$

The measurement is usually taken by adjusting the channel 1 trace to display a complete period of the reference waveform in 10 "boxes" of the scope display. This normally causes the channel 1 display to occupy the full width of the CRT. The uncertainty of the leading edge is then read in "boxes." The percent jitter is then given by

$$\% \text{ jitter} = \frac{\tau(\text{in boxes})}{10} \times 100\%$$

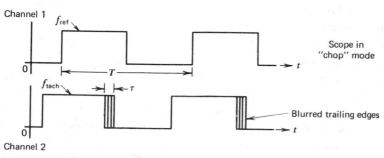

Figure 7-2 The tachometer waveform edges are not exactly periodic. Only the "average" edge is exactly periodic.

Note that calibration in units of time is not necessary for this measurement, thus eliminating the scope timebase as a source of error. Typically jitter for a single pickoff is on the order of 10% for a 1000 line disc and 12% for a 5000 line disc.

The scope is perhaps the most convenient, most graphic, and most easily understood means of knowing what the loop is doing. However, the measurement of jitter is not entirely a clearcut procedure, because it is possible to confuse jitter with oscillation of the servo. In other words, oscillation of the phase, and jitter, can look alike on the scope! Whether or not the system is oscillating can be quickly determined by monitoring the error signal to the transconductance amplifier. If the servo is oscillating, it will be apparent as a periodic component on this error signal.

FLUTTER

Another means of quantitatively measuring servo performance is with a flutter bridge. Flutter may be defined as the change in frequency $\Delta\omega$ per a nominal frequency ω. This is analogous to the jitter τ in a period T and represents the FM component of the tachometer signal:

$$\text{flutter} = K\frac{\Delta\omega}{\omega}$$

Flutter is a quantitative measurement of the FM of the tachometer feedback signal. It is measured by what is essentially an FM discriminator. By consistently using one measurement technique, or even one type of flutter meter (a number of these instruments are available commercially), flutter measurements can yield meaningful comparisons between motors and systems. To obtain such a measurement, the tachometer is connected to the input of the flutter bridge. If necessary, the speed of the motor is adjusted to conform to one of the "standard" input frequencies of the bridge (this corresponds to standard tape speeds). The meter reads flutter directly.

It is often possible to adjust system parameters (i.e., G_1, τ_M, τ_f, k_i, and k_ρ) to yield a minimum flutter. This is done by adjusting the parameters while observing the flutter bridge, noting the effect of the controls, and adjusting for minimum. Typically, a high inertia, high speed servo yields the best (lowest) flutter, all other factors being equal. The reason is that a high inertia servo, running at high speed (>0.1 oz-in-sec^2, >1000 rpm), has the least tendency to change speed because of the flywheel effect.

TORSIONAL RESONANCE

The placement of inertia can have a rather dramatic effect on flutter (Fig. 7-3). If the inertia is mounted at the end of a relatively long, thin motor shaft, it is possible, and even probable, that the system will oscillate! This is because additional poles are brought into the baseband of the servo by such an arrangement.

Consider Fig. 7-3a. With the motor running at constant speed, points X and Y on the surface of the shaft can be connected by a line parallel to the centerline of the shaft. When the motor accelerates in a CW direction, the inertia acts to retard the CW motion at the end by the shaft, and it bends or "winds up" under the resulting stress. Eventually, the speed of the inertial load "catches up" to the speed of the motor armature. However, the shaft tends to "unwind," producing an oscillatory motion. The phenomenon is called "torsional resonance." It is the result of energy storage in the shaft itself, which acts as a spring.

Under "normal" circumstances, the resonant frequency of the shaft is quite high. It is well outside of the servo passband and has no deleterious effect on performance. When a large inertia is mounted far from the motor (a few inches) the torsional resonant frequency is lowered and may fall inside the servo passband. This spells disaster for the servo. It will not lock up in phase with the reference. The cure is to avoid the situation in the first place. Notch filters or extra compensation are possible, though complicated. The best remedy is to have a thick, short motor shaft with the inertial load as close as possible to the motor.

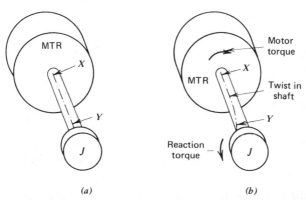

(a)   (b)

Figure 7-3 (a) At constant speed, the motor shaft is undistorted. (b) Under acceleration, the shaft stores energy as a torsional spring.

Another often encountered problem is the inability to lock up large inertial loads running at high speeds and with high density optic discs. The inertia then takes over the control of the servo. The rotational velocity cannot be adjusted quickly because of the large mass. Yet each of the disc edges demands some sort of correction, be it for jitter or actual velocity discrepancies. The two demands are incompatible, leading to the inability to establish lock. The problem can be solved by dividing the tachometer frequency before it enters the system. This produces the effect of a lower density disc and allows lockup. A 10 to 1 division is a good starting ratio, although 4:1 is often quite satisfactory. Of course, the input reference frequency must be simultaneously divided by the same number to produce the desired speed. When low speed operation from the servo is desired, the division may be switched out and the "normal" tachometer signal is used.

It may be necessary to synchronize the speeds of two or more motors. This presents no special difficulty unless it is simultaneously necessary to produce a specific angular relationship between two (or more) shafts. To establish such an angular relationship, an absolute indication of the position of both shafts is needed. This is done by including a "once-around" pulse on the optics disc. Once-around pulses may then be used to develop a phase error signal for one of the motors, causing it to align the relative positions of the shaft.

An empirical procedure for adjusting G_1, τ_M, τ_f, k_i, and k_ρ has been developed. It allows the user to adjust and stabilize the PLL virtually without doing any analytic work or even running the programs! It is based on the following facts:

1 A frequency locked loop can be produced with $k_\rho = 0$, thus eliminating the need to adjust k_ρ until the very end of the procedure.
2 When the servo loop breaks into oscillation, it emits a loud sound. Furthermore, the actual damping of the servo is readily observed on the scope by looking at the error input to the transconductance amplifier.

The procedure is as follows:

1 Adjust all controls (G_1, τ_M, τ_f, k_i, k_ρ) to zero.
2 Make sure that the motor case is grounded and the load inertia is as close as possible to the motor.
3 Adjust the reference frequency to have a midrange value. Typically this corresponds to a motor speed between 100 and 1000 rpm.
4 Advance G_1 until the system just breaks into oscillation. This is detectable

when the servo emits a sound much louder than the normal running motor noise.

5 Advance τ_f to eliminate the noise. Do not advance τ_f more than about 25% of its full range. If the noise does not disappear, readjust G_1 to a lower value. Aim to have G_1 as far advanced as possible, without advancing τ_f more than 25% of maximum. This is intended to give maximum rate damping with minimum lag.

6 Use τ_M to supplement the effect of τ_f. Do not advance τ_M more than 25% of maximum.

7 Check for a stable velocity servo. This is done by lightly grabbing the motor shaft and feeling it to find out whether it attempts to overcome the retarding torque. Switch the direction switch back and forth. The motor should change direction and be nonoscillatory in both directions.

8 Once a stable velocity servo has been obtained, slowly advance k_i and observe the channel 2 trace on the scope. At some point, the channel 2 waveform will "snap" into alignment with the channel 1 waveform. The k_i control should be advanced the minimum amount necessary for frequency lock to occur. It should lock equally well in both directions of rotation. Test this with the direction switch.

If it is not possible to obtain this frequency locked condition, the following remedies may be helpful:

1 Reduce the settings of τ_f and τ_M and accept the resulting lower setting of G_1. In other words, reduce the damping (controlled by G_1) and reduce the lag (controlled by τ_f and τ_M). In some cases it may be necessary to have $\tau_f = \tau_M = 0$.

2 Check the position of the inertia on the shaft. Move it closer to the motor.

3 If using high inertia at high speed, try dividing the tachometer signal.

4 Ground the motor and amplifier cases. It should be possible in almost all cases to obtain a frequency lock.

5 Having obtained a frequency lock, advance the k_ρ control to obtain phase-lock. The channel 2 display should align itself 180° out of phase with the channel 1 display. This separates the two leading edges of the tachometer and reference waveform. Use the maximum amount of "k_ρ" that will allow bi-directional lockup and stable operation.

Since this procedure is based on the block diagram, it can be used with many circuit variations.

Review of the Stability Problem in Control Systems

This appendix briefly and informally reviews control theory and emphasizes certain aspects of the theory that are not usually highlighted. In particular, the relationship between the Nyquist plot and Bode plot is developed. For a formal or more basic introduction to control systems, the reader is referred to any of the works listed at the end of this book.

THE TRANSFER FUNCTION METHOD

Control systems are often described in terms of block diagrams and associated transfer functions. For example, a DC motor has a block diagram and transfer function as shown in Fig. A-1 and eq. A-1. The transfer function of a permanent magnet DC motor is:

$$\frac{\omega}{I} = \frac{K_T}{Js + K_D} \tag{A-1}$$

where ω = speed (rad/sec)
I = current (A)
K_T = torque constant (oz-in./A)
J = inertia (oz-in-sec^2)
s = Laplace operator
K_D = damping constant (oz-in./rad/sec)

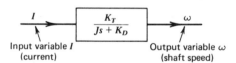

Input variable I
(current)

Output variable ω
(shaft speed)

Figure A-1   Block diagram of a permanent magnet DC motor.

The motor is viewed as a sort of "conversion module" that converts an input current I (A) to an output shaft speed ω(rad/sec). It is viewed as giving a certain "ω per I." As another example, the monostable F/V converter described in Chapter 3 has the transfer function

$$\frac{V}{\omega} = \frac{k_m}{sT_M + 1} \tag{A-2}$$

126

where V = output DC voltage (V)

ω = input frequency (rad/sec)

k_m = conversion constant (V/rad/sec)

τ_M = filter time constant (sec)

s = Laplace operator

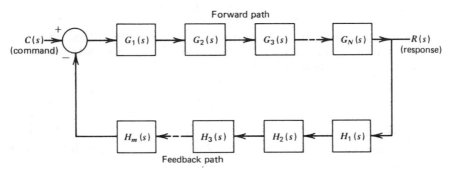

Figure A-2 The forward path and feedback path of a typical simple servo system.

The F/V converter is viewed as providing "V(volts) output per ω(rad/sec) input." In an analysis or design, all portions of the servo system are converted to block diagram form and transfer functions for the elements are established. It is then necessary to arrange the individual block diagrams into an overall system as specified or implied by the original concept of the system. A typical servo system, shown in Fig. A-2, has a "forward transfer path" consisting of $G_1, G_2 \cdots G_n$ and a "feedback path" consisting of $H_1, H_2 \cdots H_m$. C is considered to be the "command" signal and R the "response" signal. In a typical velocity servo, as described in Chapter 4, C is the input frequency and R is the output shaft speed. For the general system of Fig. A-2, the block diagram can be simplified as in Fig. A-3. $G(s)$ is called the forward transfer function and $H(s)$ the feedback transfer function. $G(s)$ is obtained by multiplying the individual transfer functions in the forward path:

$$G(s) = G_1 G_2 G_3 \cdots G_n \qquad (\text{A-3})$$

$H(s)$ is obtained by multiplying the individual transfer functions in the feedback path:

$$H(s) = H_1 H_2 \cdots H_m \qquad (\text{A-4})$$

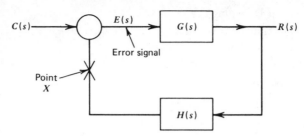

Figure A-3 A simplified block diagram showing point X. When the path is opened at point X, and excitation is applied at $C(s)$, the open loop response is measured at the output of $H(s)$.

If the loop is opened at point X, as shown in Fig. A-3, the overall transfer function between C and the output of $H(s)$ is called the open loop transfer function.

$$\text{open loop transfer function } G(s)H(s) = GH(s) \qquad \text{(A-5)}$$

The open loop transfer function has great significance in control system theory because it contains sufficient information to determine *closed* loop stability. To measure $GH(j\omega)$, it is only necessary to introduce a sine wave input at C and observe the results at X. By sweeping the input frequency, the gain and phase of the output can be measured, thereby measuring $GH(j\omega)$. $G(s)$ can usually be inferred from $G(j\omega)$. When the loop is closed at point X, the equations describing the system are

$$E(s)G(s) = R(s) \qquad \text{(A-6)}$$

$$C(s) - R(s)H(s) = E(s) \qquad \text{(A-7)}$$

Eliminating E in these equations yields

$$\frac{R}{C} = \frac{G}{1 + GH} = \text{closed loop transfer function} \qquad \text{(A-8)}$$

This is the closed loop transfer function for the system of Fig. A-3. It predicts a response R per input command C.

CRITERION FOR STABILITY

Equation A-8 can be rewritten as

$$R(s)[1 + GH(s)] = C(s)G(s) \qquad (A\text{-}9)$$

It represents the Laplace transform of a time domain differential equation describing the system. It may again be rewritten as

$$R(s)[1 + GH(s)] = C(s)G(s) + 0 \qquad (A\text{-}10)$$

The quantity 0 (zero) has been explicitly added to the right side of the equation to suggest that the solution for eq. A-11

$$R(s)[1 + GH(s)] = 0 \qquad (A\text{-}11)$$

is also a solution for eq. A-9. Values of s that satisfy eq. A-11 will automatically satisfy eq. A-9. Equation A-11 is called the *characteristic equation* of the system of Fig. A-3.

Using standard transform methods, eq. A-9 can be solved for $R(s)$ [given $C(s)$] and a total solution for $R(s)$ can be determined. Such a solution will consist of both the transient response and the forced response. (Recall that the transient response is primarily system dependent, while the forced response depends primarily on the mathematical form of the source waveform.) The control engineer most often wishes to examine first the properties of the system rather than its forced response to some particular source. To do this, it is necessary only to solve for s in eq. A-11, the characteristic equation.

As any algebraic equation, eq. A-11 has roots of the form

$$s = \sigma + j\omega \qquad (A\text{-}12)$$

where $\sigma = \mathrm{Re}\, s$
$\omega = \mathrm{Im}\, s$

In general, the roots of eq. A-11 have both real (σ) and imaginary ($j\omega$) components. The roots with imaginary components always exist in complex conjugate pairs. (This is because the coefficients of eq. A-11 are always real for a realizable servo.) In some cases, $\omega = 0$, so that the roots consist only of a real part.

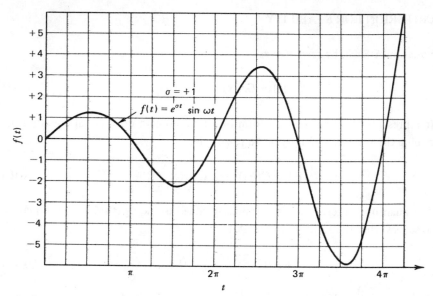

Figure A-4 An unstable system ($\sigma > 0$) is characterized by an ever-increasing output.

When the inverse Laplace transform is used to obtain the time domain solution, the typical form of the solution in control system problems is:

$$f(t) = e^{\sigma t} \sin(\omega t + \phi) \qquad \text{(A-13)}$$

An important observation can be made at this point. If $\sigma > 0$, the time domain response will have an ever expanding "envelope," or increase indefinitely in time. The condition of $\sigma > 0$ constitutes an unstable condition and is illustrated in Fig. A-4. If $\sigma < 0$, the time domain response will have a decaying envelope. The transient response will decay in time and eventually disappear after some number of time constants. This constitutes a stable condition and is illustrated in Fig. A-5. The condition $\sigma = 0$ is considered unstable because it represents a sustained oscillatory condition with no damping.

Pure exponential roots ($\omega = 0$) obey the same rules. When $\sigma > 0$, the exponential increases indefinitely in time and the output response increases until it reaches some physical limit. Such a system is unstable. For $\sigma < 0$, the transient decays in time and the system is considered to be stable. The case with $\sigma = 0$ is considered unstable because the transient never decays.

The conclusion is that for stability the real part of the root must be negative.

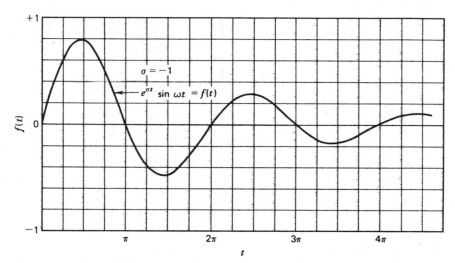

Figure A-5 A stable system ($\sigma < 0$) is characterized by a decaying transient that eventually reaches zero.

Stated another way, there may be no solutions of eq. A-11 in the right half s plane. (If there were solutions in the right half s plane, their real parts would be positive.) This may be represented graphically as shown in Fig. A-6.

THE DIRECT METHOD

A central problem in control theory is to determine whether there are any roots of the characteristic equation in the RHP. A second problem is how to change the characteristic equation to avoid such roots.

One way to determine stability is to find the roots of the characteristic equation directly. To do this, it is necessary to know the open loop transfer function. As discussed earlier, the open loop transfer function can be measured. However, since the transfer functions of the individual elements are usually available, the open loop transfer function can be evaluated from eqs. A-3 and A-4. Assume that the open loop transfer function has the general form of eq. A-14:

$$GH(s) = K\frac{(s + z_1)(s + z_2)(s + z_3)\cdots}{(s + p_1)(s + p_2)(s + p_3)\cdots} \qquad (A\text{-}14)$$

Here z_n and p_n may be complex numbers. The characteristic equation may then be written as

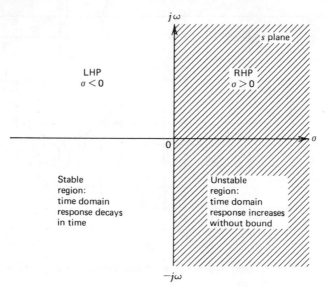

Figure A-6 Roots of a stable servo must lie in the left half s plane (LHP).

$$K\frac{(s+z_1)(s+z_2)(s+z_3)\cdots}{(s+p_1)(s+p_2)(s+p_3)\cdots}+1=0 \qquad (\text{A-15})$$

To illustrate the method, assume for the moment that GH consists of only the first three terms of the numerator and denominator; eq. A-15 can be expanded to yield

$$(K+1)s^3+s^2\left[K(z_1+z_2+z_3)+(p_1+p_2+p_3)\right]+s[K(z_1z_2+z_1z_3+z_2z_3)$$

$$+(p_1p_2+p_1p_3+p_2p_3)]+Kz_1z_2z_3+p_1p_2p_3=0 \qquad (\text{A-16})$$

This equation results from a relatively simple third order system. (The "order" of a system refers to the highest power of s in the expanded denominator of $GH(s)$ such as in eq. A-16). It is necessary to know whether this equation, which is a form of the characteristic equation, has any roots in the RHP.

Since z_1, z_2, z_3, p_1, p_2, p_3, and K are known, it is entirely possible to obtain a numerical solution for the three roots of this equation. However, if these solutions include a value of s such that $\text{Re}(s) \geqslant 0$, there is no procedure, or guide, to modify the system and move the roots to the LHP. The direct method of actually finding the roots of the characteristic equation lends no insight into how to modify the system for stability. The engineer is left to guess which of the parameters (seven in this case) to adjust and how to adjust them. For this reason, the direct method is abandoned.

THE NYQUIST STABILITY CRITERION

The saving grace is that it is generally not necessary to know the root, but only to be sure that it is not in the RHP. Such a determination is made by using the Nyquist criterion for stability. The Nyquist criterion is used to determine *closed* loop servo stability from the *open* loop transfer function. The method is presented here without proof, since many excellent treatments of the Nyquist criterion exist.

It is first assumed herein that the *open* loop transfer function $GH(s)$ has no poles in the right half of the s plane. This implies that the *open* loop system is stable. For the motor control phaselock loop, this is always true. The Nyquist criterion is used to detect the presence of RHP zeros of the closed loop characteristic equation, using only the open loop transfer function.

The test devised by Nyquist depends on the property of "conformal mapping" (Fig. A-7). For example, conformal mapping implies that if a certain region is encircled in the s plane, another corresponding region will be encircled in the $F(s)$ plane. The encirclement in the s plane is "mapped" into the $F(s)$ plane. This is a theorem in the mathematics complex variables and will not be proved. It is sufficient to understand that if the contour C in the s plane is closed, then the contour C' in the $F(s)$ plane is also closed. Assume now that only a single zero of $F(s)$ exists inside the contour C. Then it can be proved that contour C' [in the $F(s)$ plane] performs a single *clockwise* encirclement of the origin. Furthermore, if only a single *pole* exists inside of contour C, then contour C' performs a single *counter clockwise* rotation about the origin.

In general, the number of clockwise rotations of C' about the origin [in the $F(s)$ plane] is equal to the difference between the number of zeros and number of poles inside contour C in the s plane:

$$\text{number of clockwise rotations of } C' \text{ about origin} = Z - P \qquad \text{(A-17)}$$

where Z = number of zeros inside C
$\quad\ P$ = number of poles inside C

The problem at hand is to determine whether there are zeros of $F(s) = 1 + GH(s)$ *anywhere* in the entire RHP. Therefore, the contour C should enclose the *entire*

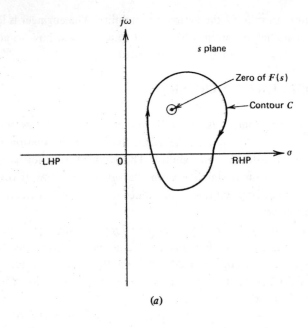

(a)

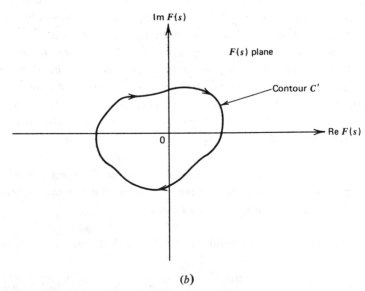

(b)

Figure A-7 (*a*) A closed contour *C* in the *s* plane that encloses a zero of *F*(*s*). (*b*) Contour *C′* in the *F*(*s*) plane is generated by using values of *s* on contour *C* in the *s* plane. If *C* is closed, *C′* will be closed. Note that, since *C* encloses a zero, *C′* encircles the origin.

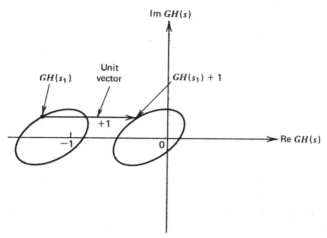

Figure A-8 If $GH(s)$ encircles the -1 point, then $GH(s) + 1$ encircles the origin.

right half s plane. $F(s) = 1 + GH(s)$ will encircle the origin once (clockwise) for every *zero* in the RHP and counterclockwise for every *pole* in the RHP.

Note that if $GH(s)$ encircles the -1 point, then $GH(s) + 1$ will encircle the origin (Fig. A-8). This is because the addition of quantity "1" to $GH(s)$ moves every point of $GH(s)$ one unit to the right. Consequently, it is adequate to determine how many times (if any) $GH(s)$ encircles the -1 point. This is equivalent to $GH + 1 = F(s)$ encircling the origin.

To have contour C encircle the entire right half s plane, it is common (but not necessary) to use the so-called Bromwich contour for C. Consider Fig. A-9,

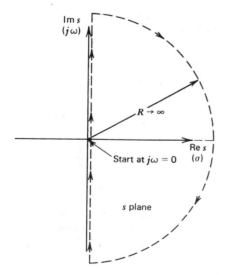

Figure A-9 The Bromwich contour encircles the entire right half s plane (RHP). The arrows show the direction of progression along the path.

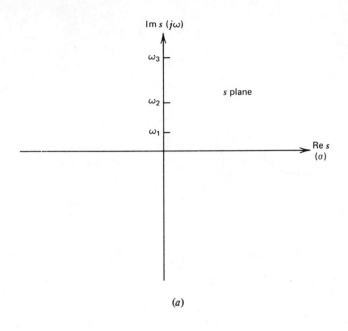

(a)

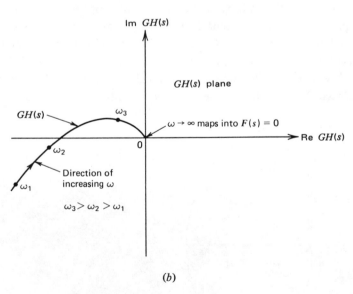

(b)

Figure A-10 (a) As ω increases toward $+\infty$ along the Bromwich contour, $GH(s)$ decreases toward zero in the $GH(s)$ plane. (b) Showing $GH(s)$ approaching zero as $\omega \to \infty$.

136

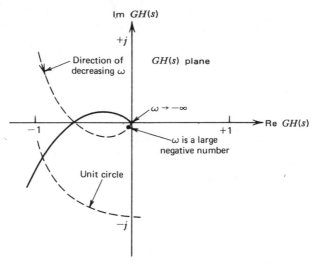

Figure A-11 The dotted line shows $GH(s)$ for $-\infty < \omega \le 0$. The dotted line is a mirror image of the solid line, which is a plot of $GH(s)$ for $0 \le \omega < +\infty$. Note that the -1 point is not encircled, indicating that the system is stable.

which shows the Bromwich contour starting at the origin of the s plane and following the $j\omega$ axis toward infinity. At infinite distance, it encircles the entire RHP, until it reaches the negative portion of the imaginary axis. It then follows the $j\omega$ axis from $-\infty$ back toward the origin. If $GH(s)$ has singularities on the $j\omega$ axis (such as a pure integration which is a pole at $s = 0$), these are skirted by modifying the Bromwich contour.

For the purposes of this review, it is important to emphasize only the following facts about the Nyquist plot when used with the motor control equations developed here:

1 As the frequencies along the $j\omega$ axis, starting from 0 and moving toward infinity (i.e., along the Bromwich contour), are substituted into $GH(s)$ (see Fig. A-10a), $GH(s)$ takes on a general shape shown in Fig. A-10b. Each frequency along the $j\omega$ axis produces a new point in the $GH(s)$ plane.

2 The point $\omega \to \infty$ maps into the point $GH(s) = 0$.

3 As the frequencies along the $-j\omega$ axis from $\omega \to -\infty$ toward $\omega = 0$ are substituted into $GH(s)$, a mirror image of the plot of Fig. A-10a develops. This is illustrated in Fig. A-11. Note that this reflects a stable servo because the -1 point is not encircled.

4 If the system were unstable, it would appear as in Fig. A-12.

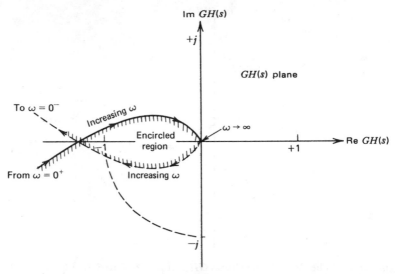

Figure A-12 One way to determine whether the −1 point is encircled is to shade the region to the right of the $GH(s)$ plot with respect to the direction of increasing ω. Note that ω *increases* as it proceeds from −∞ to zero. The −1 point is encircled in this case.

The following conclusion can be reached from an examination of the Nyquist plot. If the −1 point is encircled by $GH(s)$, there is a zero of $1 + GH(s)$ in the RHP and the servo is unstable. If the −1 point is not encircled, there are no zeros of $1 + GH(s)$ in the RHP and the servo is stable. Recall that $GH(s)$ is assumed to be *open loop* stable and therefore has $GH(s)$ no poles in the RHP. Consequently any *CW* encirclements of −1 will indicate the presence of zeros. To establish stability, it is only necessary to plot $GH(j\omega)$ in the immediate region of the −1 point. The usual practice is to draw $GH(s)$ only for positive values of ω. The other half is simply the mirror image of the first half and is not usually drawn, but inferred. Note that *open loop* data $GH(s)$ has been used to predict *closed loop* stability.

RELATIVE STABILITY

Although it is necessary and important to determine the stability of a servo, up to this point it is a "yes-no" test. By itself, it would not be overwhelmingly useful were it not for the idea of "relative stability." The system of Fig. A-13a is unstable because the −1 point is encircled. The systems in Fig. A-13b and c are

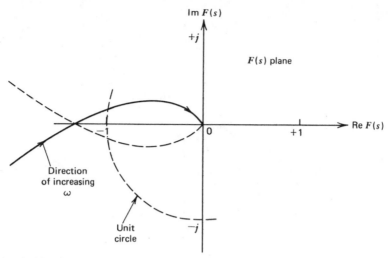

Figure A-13a Nyquist plot for an unstable servo system. Note that the −1 point is encircled.

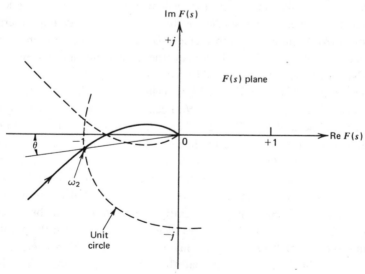

Figure A-13b Nyquist plot for a stable servo system with a "small" phase margin "θ."

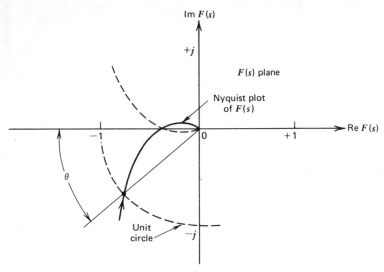

Figure A-13c Nyquist plot for a stable servo system with a "large" phase margin "θ."

stable because the -1 point is not encircled. But what is the difference between the systems in Fig. A-13b and A-13c? According to the Nyquist criterion, both are stable. The difference is that the system in Fig. A-13c will have less overshoot, less "ringing," and a smaller settling time than the system described by the plot in (b). System (c) is considered *relatively* more stable than system (b). This factor of relative stability is perhaps the most significant and useful result of the Nyquist plot, and is discussed below.

Consider the sketch of Fig. A-14, showing a typical Nyquist diagram for a stable servo. The complex number $GH(s)$ can be represented by an arrow from the origin to the plot of $GH(s)$. Furthermore, the unit quantity is represented by a unit length arrow pointing to the right. From the diagram it is evident that the quantity $1 + GH(s)$ can be represented by the arrow from the tip of the unit vector to the origin. By simply translating that arrow $GH(s) + 1$ in space, so that it ends at the -1 point, it becomes clear that the distance from the -1 point to the $GH(s)$ curve is the quantity $1 + GH(s)$. It can be observed in Fig. A-14 that as the angle θ decreases (θ is the angle between the negative real axis and the $GH(s)$ "vector"), the quantity $1 + GH(s)$ also decreases. When $\theta = 0$, $1 + GH(s) = 0$. Recall that the closed loop response has the form

$$\frac{R}{C} = \frac{G}{1 + GH} \qquad (A\text{-}8)$$

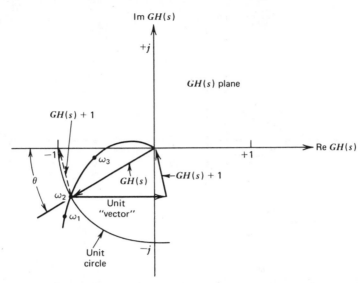

Figure A-14 The quantity $GH(s) + 1$ is represented by the "vector" originating at the tip of the $GH(s)$ "vector" and ending at the -1 point.

As the angle θ decreases, the magnitude of denominator of the transfer function decreases, and the "response per input command" takes on ever-increasing values. When the denominator $[=1 + GH(s)]$ takes on very small values (θ small), a large response R is produced for even a very small command C. This is illustrated in Fig. A-13b, where θ is small. A corresponding *closed loop* frequency response plot is sketched in Fig. A-15 to show the peaking effect. As θ decreases, the peaking near ω_2 increases because the numerator of the

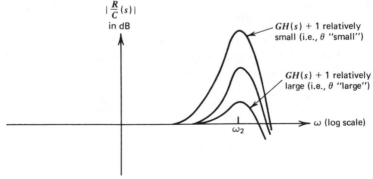

Figure A-15 Closed loop frequency response for several (relative) values of θ.

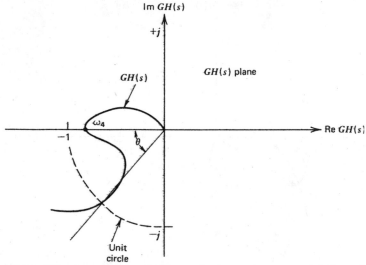

Figure A-16 Nyquist plot of a highly underdamped system with "good" phase margin. $GH(s) + 1$ is very small near $\omega = \omega_4$.

closed loop transfer function becomes small. Severe peaking of the frequency response curve implies a ringing, underdamped, almost oscillatory system.

The importance of controlling θ has led to giving θ a name: the "phase margin." But controlling the phase margin ("PM") does not necessarily always ensure a stable, adequately damped servo, as will now be shown.

Consider the Nyquist plot of Fig. A-16. The phase margin is quite large, yet at ω_4 the quantity $1 + GH(j\omega_4)$ is quite small. To control the situation depicted in Fig. A-16, it is necessary to control the magnitude of $GH(j\omega)$ when its angle is $180°$. In other words the magnitude of $GH(j\omega)$ must be substantially less than unity when its angle is $180°$. Because of the importance of the gain magnitude at an angle of $180°$, it is given a name: the "gain margin" (in dB).

To obtain an adequately damped system, it is necessary to control at least two characteristics of the Nyquist plot: the phase margin and the gain margin. These are defined as shown in Fig. A-17.

$$\text{phase margin} = \angle GH(j\omega)\big|_{\substack{\text{at}\\ \text{crossover}\\ \text{of unit}\\ \text{circle}}} + 180° \tag{A-18}$$

$$\text{gain margin} = 20\log\left|\frac{1}{GH(j\omega_G)}\right| \tag{A-19}$$

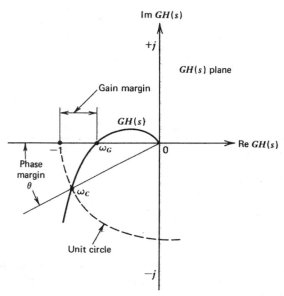

Figure A-17 Phase margin θ (in degrees) is defined as the angle between the negative real axis and the line joining the origin with the point where $GH(j\omega)$ intersects the unit circle. The frequency at which $GH(j\omega)$ intersects the unit circle is called the crossover frequency ω_C. The gain margin is defined as the number by which to multiply $GH(j\omega_G)$ to bring the product $GH(j\omega_G) \times$ gain margin to unity. The frequency at which $GH(j\omega)$ crosses the negative real axis is called ω_G. Gain margin can be expressed in dB.

ω_G is the frequency at which $GH(s)$ crosses the real axis closest to the -1 point. The purpose of controlling these quantities is to ensure that the quantity $1 + GH(j\omega)$ stays relatively large compared to $G(j\omega)$ and thus avoids excessive peaking of the closed loop transfer function.

A system can have "good" phase and gain margins, yet be excessively under-damped. This is illustrated in Fig. A-18. At ω_4, the denominator of $1 + GH(j\omega)$ is relatively small, indicating a strong peak in the closed loop frequency response. Yet the margins for this system would be considered "good." The easiest way to avoid problems without defining even more figures of merit is to simply do the Nyquist plot, even if only as a check on the design made by other means. If the plot approaches the -1 point in some unexpected way, it will be immediately apparent. This is the reason for including the Nyquist plot in the computer design process (Chapter 6).

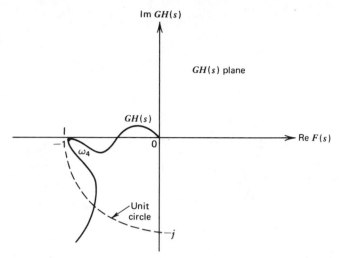

Figure A-18 A case where both phase and gain margins are "good." Yet at ω_4, a strong peak will occur in the closed loop response plot, implying severe underdamping at $\omega = \omega_4$.

THE BODE METHOD

The design and compensation of a servo system using Nyquist techniques directly is a cumbersome, laborious task. However, the Nyquist criterion is the foundation of many other techniques that provide results with much less labor and allow much greater insight into what to do to produce a stable system.

One technique, perhaps the simplest one, available for the design of servo compensation is the Bode method. The Bode method uses two graphs to plot the same data that the Nyquist plot contains in one graph. First, the Bode method plots the magnitude of $GH(j\omega)$ in dB ($= 20 \log GH(j\omega)$) versus $\log \omega$, *thereby allowing the addition of the various individual transfer functions rather than requiring multiplication as on the Nyquist diagram.* (Recall that quantities in dB can be added.) When doing the design it is much easier to see how to *add* in compensation than how to *multiply in* compensation. Second, the Bode method plots the angle of $GH(j\omega)$ as a function of frequency on a separate graph. Third, the Bode plot uses an "asymptotic approximation" for $GH(j\omega)$ rather than requiring a detailed plot as in the Nyquist criterion. This permits much more rapid construction and "testing" on the graph.

To obtain the asymptotic approximation, the open loop transfer function is put into the so called Bode form. This can always be done by manipulating variables until all terms have the general form of eq. A-20:

$$GH(j\omega) = \frac{K\left(j\dfrac{\omega}{\omega_4}+1\right)\cdots}{\left(j\dfrac{\omega}{\omega_1}+1\right)\left(j\dfrac{\omega}{\omega_2}+1\right)\left(j\dfrac{\omega}{\omega_3}+1\right)\cdots\left(j2\zeta\dfrac{\omega}{\omega_{N_1}}+\left(1-\left(\dfrac{\omega}{\omega_{N_1}}\right)^2\right)\right)}$$

(A-20)

The magnitude plot can be obtained from:

$$dB = 20\log|GH(j\omega)| = 20\log K + 20\log\left|j\frac{\omega}{\omega_4}+1\right| + 20\log\left|j\frac{\omega}{\omega_5}+1\right| + \cdots$$

$$-20\log\left|j\frac{\omega}{\omega_1}+1\right| - 20\log\left|j\frac{\omega}{\omega_2}+1\right| - \cdots \quad (A-21)$$

This equation is rewritten as eq. A-22:

$$dB = 20\log K + 20\log\sqrt{\left(\frac{\omega}{\omega_4}\right)^2+1} + 20\log\sqrt{\left(\frac{\omega}{\omega_5}\right)^2+1} + \cdots$$

$$-20\log\sqrt{\left(\frac{\omega}{\omega_1}\right)^2+1} - 20\log\sqrt{\left(\frac{\omega}{\omega_2}\right)^2+1} - \cdots \quad (A-22)$$

To illustrate how the asymptotic approximation is made for the second numerator term of the equation, consider what happens when ω is very small compared to ω_4:

$$dB = 20\log\sqrt{\left(\frac{\omega}{\omega_4}\right)^2+1} \cong 20\log 1 = 0 \text{ dB} \qquad (\text{for } \omega \ll \omega_4)$$

Furthermore, when ω is much larger than ω_4, we have

$$dB = 20\log\sqrt{\left(\frac{\omega}{\omega_4}\right)^2+1} \cong 20\log\frac{\omega}{\omega_4} \qquad (\text{for } \omega \gg \omega_4)$$

When $\omega \gg \omega_4$, there is a linear relationship between dB and log ω. These two cases are sketched in Fig. A-19. The terms "much greater than" or "much less than" are interpreted here as being an order of magnitude (i.e., a factor of 10) greater or less in the sketch of Fig. A-19. This is done only for con-

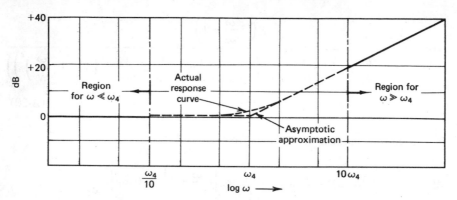

Figure A-19 Asymptotic approximation of the term $(1 + j\omega/\omega_4)$. The actual response differs from the approximation by 3 dB maximum at $\omega = \omega_4$.

venience. Note that when $\omega = 10\omega_4$, $20 \log \omega/\omega_4 = 20 \log 10\omega_4/\omega_4 = 20$ dB. A 20 dB rise exists for each order of magnitude that ω increases.

The asymptotic approximation is completed by projecting the two extreme ends of Fig. A-19 toward ω_4 in a linear fashion. The resulting plot is shown in Fig. A-19. If an exact plot is made, the worst case error (between the exact plot and the asymptotic approximation) occurs at $\omega = \omega_4$. The approximation shows that at $\omega = \omega_4$ the magnitude is 0 dB. However, the actual curve at $\omega = \omega_4$ is evaluated as

$$ dB = 20 \log \sqrt{\left(\frac{\omega}{\omega_4}\right)^2 + 1} = 20 \log \sqrt{1 + 1} = 3 \text{ dB} $$

This shows the maximum error to be 3 dB.

Quadratic terms are evaluated similarly, except that they can differ very markedly from the asymptotes at $\omega = \omega_N$. This depends on the value of ζ, the "damping factor" of the quadratic term. The magnitude of a quadratic term is plotted in Fig. E-1 (Appendix E). Note that near $\omega = \omega_N$ the actual curve can differ substantially from the asymptotic approximation, depending on the value of ζ.

When two terms are to be multiplied in $GH(j\omega)$, their Bode plots must merely be added. For example, assume a $GH(j\omega)$ having a form such as in eq. A-23. The individual plots are shown as dotted lines. The overall response curve is shown as solid lines and is simply the addition of the two dotted curves (see Fig. A-20). Equation A-23 is plotted in the solid line:

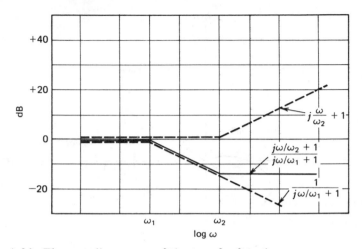

Figure A-20 The overall response of the transfer function
$$F(j\omega) = [j(\omega/\omega_2) + 1]/[j(\omega/\omega_1) + 1]$$
is obtained by adding the asymptotic approximations of the component terms.

$$F(j\omega) = \frac{j\dfrac{\omega}{\omega_2} + 1}{j\dfrac{\omega}{\omega_1} + 1} \qquad\qquad\text{(A-23)}$$

The general practice in designing a servo with the Bode method is to simply plot the asymptotic approximation for $GH(j\omega)$ and *control the phase angle at the* 0 dB *crossover point.* Generally it is not necessary to plot the angle of $GH(j\omega)$, especially if the Nyquist plot is later used to check the results. Stated another way, the Bode design method seeks to control the angle of $GH(j\omega)$ at its 0 dB crossover so that the desired phase margin is produced. This is usually done by having the Bode plot (i.e., the asymptotic approximation) cross the 0 dB axis at a slope of -20 dB/decade. This is perhaps the easiest design procedure available to the engineer. But why does it work? Why is only the 0 dB crossing of significance? Why cross at a slope of -20 dB/decade? Only when the Bode method is considered in the light of the Nyquist criterion (from which it was derived) do these rules begin to make sense.

The first thing that the designer attempts to do to ensure stability is to have an adequate phase margin on the Nyquist plot. The phase margin was defined as the angle of $GH(j\omega)$ *at the unit circle crossover* (see Fig. A-21a). It is essential to realize clearly that the phase margin is defined *on the unit circle.*

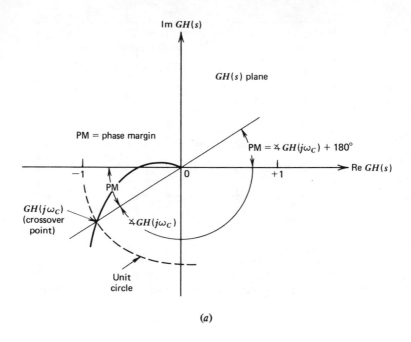

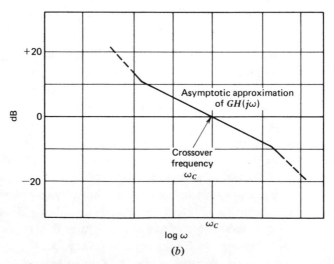

Figure A-21 (*a*) The phase margin (PM) is defined at the unit circle crossover point with $GH(s)$. Note that $\angle GH(j\omega_C) + 180° = $ PM. (*b*) The 0 dB crossover of the asymptotic approximation corresponds to the unit circle crossover of the Nyquist plot. (0 dB corresponds to a gain of 1.)

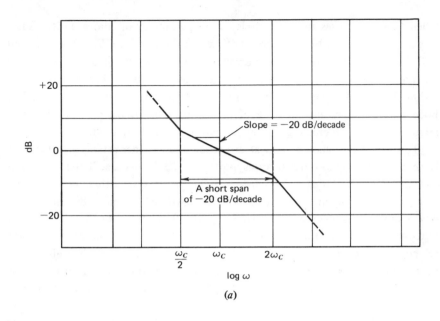

(a)

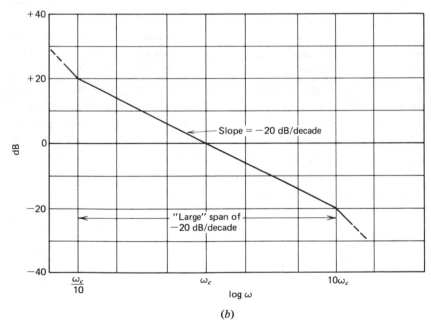

(b)

Figure A-22 (*a*) A short span of −20 dB/decade slope about crossover implies a relatively small phase margin. (*b*) A long span of −20 dB/decade slope about crossover implies a large phase margin.

149

When $GH(j\omega)$ crosses the unit circle, its magnitude is, of course, unity. But a magnitude of unity is by definition 0 dB! When the angle of $GH(j\omega)$ is evaluated at 0 dB, on the Bode plot, it is equivalent to evaluating the angle of $GH(j\omega)$ on the unit circle on the Nyquist plot. The phase margin is then calculated by adding 180° to the angle of $GH(j\omega)$ (at 0 dB crossover) (see Fig. A-21a). The Bode plot is simply another way of presenting the Nyquist plot. It is for this reason that the 0 dB crossing is so important in the Bode method.

To control the phase margin at crossover, the designer attempts to have the Bode plot of $GH(j\omega)$ cross at an asymptotic -20 dB/decade. If the Bode plot crosses the 0 dB axis at -20 dB/decade, it can be shown that the *maximum* phase shift (i.e., the maximum angle of $GH(j\omega)$) at crossover is 180°, no matter how many other poles exist in $GH(j\omega)$. By forcing the open loop transfer function to cross the 0 dB axis at -20 dB/decade, the phase margin will *always* be positive (i.e., no encirclement of the -1 point with the Nyquist plot). Furthermore, the greater the span of -20 dB/decade (how far above and below crossover the slope continues to be -20 dB/decade), the greater the phase margin will be. (See Fig. A-22.)

Controlling the span of -20 dB/decade at the 0 dB crossover of the Bode plot is a simple means of controlling the phase margin. The span is adjusted to create the desired phase margin. For this reason a slope of -20 dB/decade is "forced" at the 0 dB crossover. By so controlling the phase margin, the resulting system will almost always provide satisfactory transient performance. However, as previously mentioned, a "good" phase margin is in general insufficient to guarantee stability, even though in many situations a good phase margin proves to be adequate. To be sure, it is always prudent to do the Nyquist plot as a check on the results. The programs provided in Chapter 6 perform the design based on controlling the span of the -20 dB/decade slope of a Bode plot. The design is then checked by way of a Nyquist plot and also a closed loop response plot. This leaves no room for doubt!

B

Position, Velocity, and Phaselock Servos

THE PLL AS A MOVING POSITION SERVO

The motor control servo is in many ways analogous to the conventional position servo shown in Fig. B-1. Conventional servos have been extensively analyzed in the literature, and this body of knowledge can be used to guide the analysis and improve the understanding of the PLL.

This time honored positioning loop can be briefly explained as follows. In Fig. B-1 pot A develops a command signal θ_{ref} proportional to the angular rotation of the pot. Pot B develops a signal θ_{out} proportional to the angular position of the motor shaft. (The tachometer, shown as dotted lines, is discussed later.) When an angular difference between θ_{ref} and θ_{tach} exists, error voltage ϵ is produced by the error detector. The error is proportional to the difference between the angles. The error signal is amplified in A_1 and used to drive the motor, which in turn drives pot B to reduce the error between θ_{ref} and θ_{out}. As the difference between the angles and the resulting signal to the motor become very small, motor friction retards further rotation. At that point the angular difference is small and the final position has been attained.

With slightly different semantics, the input angle can be called the reference phase, the output angle the output phase, and the error detector called the phase detector. To develop the analogy between the conventional position loop and the PLL, it is necessary to replace only the mechanism of the transducers. The

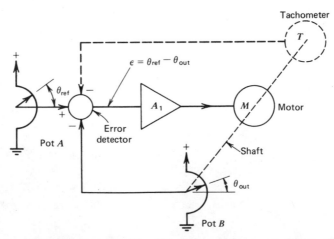

Figure B-1 Block diagram of a "traditional" position servo. The position transducer is potentiometer B, which converts the shaft angle θ_{out} to a voltage proportional to angular rotation. The tachometer is often necessary to stabilize the position loop.

152

concepts are identical. The input signal can be replaced by a periodic waveform with a phase θ_{ref}. The output signal, instead of being derived from a pot, is derived from an optical tachometer. At lock, it will have a phase θ_{tach} (instead of θ_{out}). The error detector is replaced by a phase detector, which provides an output voltage proportional to (θ_{ref} − θ_{tach}). Appropriate phase detectors are described in Chapter 3. The motor then drives the optical tachometer to reduce the phase error between the signals to almost zero. The block diagram of a simple motor control PLL is shown in Fig. B-2.

Whenever the reference signal and tachometer signal are out of phase, error voltage ϵ increases to drive the motor and reduce the phase difference. When the phases are almost aligned, a small error remains to provide drive to the motor. The equations written to describe these two systems are identical! In the first case, the shaft is stationary in its final state. In the case of the PLL, the shaft has a constant velocity such that $\omega_{ref} = \omega_{tach}$. For this reason, the motor control PLL is often called a "moving position servo." Since the equations for the systems are the same, it should be clear that the techniques applicable to the conventional position loop should be carefully examined for their usefulness in the PLL servo.

Two techniques are often incorporated in the conventional position loop. First, "rate damping" is included for stability. By rate damping is meant that a signal proportional to velocity is caused to be part of the error signal. Figure B-1 shows a tachometer (as dotted lines) for this purpose. Second, the use of a *transconductance amplifier* for A_1 (which provides a current output

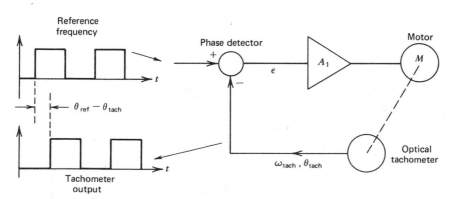

Figure B-2 Block diagram of a simple PLL. The phase detector replaces the error detector of Fig. B-1. The optical tachometer output contains both position (i.e., phase) and velocity (i.e., frequency) information, thereby replacing the position sensing pot and the tachometer of Fig. B-1.

from a voltage input) is of great advantage. To understand these points, it is helpful to analyze the conventional rate damped servo shown in Fig. B-3. The results have been applied directly to the PLL. The block diagram is derived directly from Fig. B-1, with a tachometer added in the feedback path for rate damping. Laplace transform notation is used to describe the blocks. The forward transfer function is written as

$$G(s) = \frac{A_1 G_1}{s} \qquad \text{(B-1)}$$

where G_1 = motor transfer function
A_1 = amplifier transfer function
$1/s$ = pure integration. Converts shaft speed to shaft position

The feedback loop transfer function is written as

$$H(s) = sk_t + k_p \qquad \text{(B-2)}$$

where k_ρ = position transducer gain (V/rad)
k_t = tachometer gradient (V/rad/sec)
s = Laplace operator

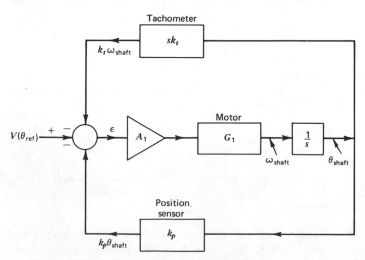

Figure B-3 Block diagram of the position servo suitable for analysis. s = Laplace operator.

The open loop transfer function is written as

$$GH(s) = \frac{A_1 G_1 k_\rho \left(s\frac{k_t}{k_p} + 1 \right)}{s} \tag{B-3}$$

The closed loop transfer function can be written as

$$\frac{\theta_{\text{shaft}}}{\theta_{\text{ref}}} = \frac{\dfrac{A_1 G_1}{s}}{1 + \dfrac{A_1 G_1 k_\rho}{s}\left(s\dfrac{k_t}{k_p} + 1 \right)} \tag{B-4}$$

To proceed with the analysis, it is necessary to know $G_1(s)$, the transfer function of the DC motor. There are two choices for G_1, depending on whether the DC motor is to be driven by a *current source* or a *voltage source*. If A_1 is constructed to be a transconductance amplifier (which provides A_1 *amperes* for every input volt), the motor is driven from a current source and its transfer function is

$$G_1(s) = \frac{\omega}{I} = \frac{K_T}{Js + K_D} \tag{C-13}$$

This is derived in Appendix C.

If A_1 is constructed as a voltage amplifier (which provides A_1 volts for every input volt), the motor transfer function becomes

$$G_1(s) = \frac{\omega}{V} = \frac{1/K_e'}{(s\tau_1 + 1)(s\tau_2 + 1)} \tag{C-18}$$

where τ_1 and τ_2 are as given in eq. C-19 and C-20. This is also derived in Appendix C.

Consider first the case where A_1 is a transconductance amplifier and the motor transfer function is given by eq. C-13. From eq. B-3, the open loop transfer function is found to be

$$GH = \frac{A_1 K_T k_p \left(s\frac{k_t}{k_p} + 1 \right)}{s(sJ + K_D)} \tag{B-5}$$

The characteristic equation of the system is

$$GH + 1 = \frac{A_1 k_p K_T \left(s\frac{k_t}{k_p} + 1\right)}{s(sJ + K_D)} + 1 = 0 \qquad \text{(B-6)}$$

By algebra:

$$s^2 + s\left(\frac{K_D + A_1 k_t K_T}{J}\right) + \frac{A_1 k_p K_T}{J} = 0 \qquad \text{(B-7)}$$

The characteristic equation can be solved for s to yield

$$s = \sigma + j\omega \qquad \text{(B-8)}$$

where

$$\sigma = -\frac{K_D + A_1 K_T k_t}{2J} \qquad \text{(B-8a)}$$

$$\omega = \sqrt{\left(\frac{K_D + A_1 K_T k_t}{2J}\right)^2 - \frac{A_1 k_p K_T}{J}} \qquad \text{(B-8b)}$$

For a well designed motor, K_D is small. This implies that without the tachometer ($k_t = 0$) the servo would be highly underdamped. However, the tachometer gradient k_t allows control of the real part of the root, thereby controlling the damping. "Rate damping" is effective in controlling the decay time of the transients. The damping and overshoot of the response can be adjusted by feeding back appropriate amounts of DC tachometer voltage. Although k_t is fixed by the tachometer design, the "effective" k_t can be varied by using a pot divider on the output of the tachometer, for example. This allows adjustment of the effective k_t. The use of rate damping and a transconductance amplifier facilitates the control of dynamic response.

If A_1 were selected to be a voltage amplifier, the open loop transfer would be

$$GH(s) = \frac{A_1 k_p \left(s\frac{k_t}{k_p} + 1\right)}{K_e s(s\tau_1 + 1)(s\tau_2 + 1)} \qquad \text{(B-9)}$$

The characteristic equation is easily shown to be a cubic equation in s. Experience with such equations indicates that an adequate phase margin is difficult to attain, especially if τ_1 and τ_2 occur at relatively low frequencies. If the numerator term were adjusted to cancel one of the denominator terms, there would often still be a phase shift of close to 180° near the crossover of a system with high A_1. The phase margin would be close to zero. That is not to say that the system cannot be stabilized, but only that it is more difficult to do so. If there were any other poles in GH (such as the carrier filter in the PLL), it would quickly become impossible to stabilize the system.

The key points are these. First, feedback of some velocity signal is necessary to control damping. Second, it is highly desirable to have A_1 as a transconductance amplifier and therefore to consider the motor to be a current operated device. The transfer function of the motor then has *one less pole* than if it were voltage driven. The pole due to the inductance (and any added inductance) simply disappears! (This is not magic, but merely a reflection of the fact that the current amplifier negates the current retarding property of the inductance. A_1 "jams" current into the motor and inductor by automatically increasing the voltage as needed).

It is for these reasons that A_1 is selected as a transconductance amplifier and a velocity dependent signal is always included in the loop. These factors are incorporated into the design of the PLL. A disadvantage of using the "moving position servo" analogy is that it implicitly demands a phase detector as the error detector. In Chapter 3 it is shown that there are no entirely satisfactory phase detectors for motor control loops. This leads to the attempt to "outflank" the need for a phase detector at all.

THE PLL COMPARED TO A
CONVENTIONAL VELOCITY SERVO

Another view of the motor control PLL that yields useful results is as a velocity servo with a pure integration of the error. Again, an analogy will be made between a conventional velocity servo using a DC tachometer and the PLL servo. The conventional velocity servo is shown in Fig. B-4. This servo operates as follows. Error voltage is proportional to the difference between the tachometer voltage and the reference voltage:

$$\epsilon = V_{ref} - V_{tach} \qquad (B\text{-}10)$$

The tachometer turns at a speed to reduce the error to a small value. However,

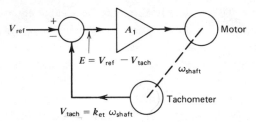

Figure B-4 Block diagram of a simple velocity servo.

some error is necessary to drive the motor, and there cannot be an exact correspondence between V_{ref} and V_{tach}.

A means of attaining zero difference between V_{ref} and V_{tach} is to insert an integrator in the forward path. Such a scheme is depicted in Fig. B-5. This servo will reach a steady state condition only when $\epsilon = 0$. The error ϵ *must* go to zero for a steady state condition to occur. This can be appreciated from an examination of the relationship between ϵ and ϵ':

$$\epsilon' = \int \epsilon \, dt + C \tag{B-11}$$

If $\epsilon \neq 0$, $\epsilon' = \epsilon t + C$ and ϵ' grows continuously in time, without bound. Only if $\epsilon = 0$ is the output of the integrator a stable constant C. This constant drives the motor when $\epsilon = 0$.

The tachometer voltage is directly proportional to tachometer speed. Furthermore, the reference voltage can be thought of as a "reference speed" being converted to a voltage by the same gradient as the tachometer. (This can be conceived of as spinning a "reference tachometer" at the desired speed, analogous to rotating a pot to the desired angle.) The equation for the error can then be rewritten as

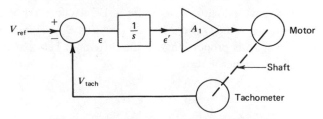

Figure B-5 Velocity servo with pure integration.

$$\epsilon = V_{\text{ref}} - V_{\text{tach}} = K_{et}(\omega_{\text{ref}} - \omega_{\text{tach}}) \tag{B-12}$$

This gives the mathematically correct result. Substituting into eq. B-11, we have

$$\epsilon' = \int K_{et}(\omega_{\text{ref}} - \omega\text{tach})dt \tag{B-13}$$

But what is the interpretation of the integral of velocity? It is the position signal, or the phase!

$$\epsilon' = K_{et}(\theta_{\text{ref}} - \theta_{\text{tach}}) + C \tag{B-14}$$

where $\int(\omega_{\text{ref}} - \omega_{\text{tach}})dt = (\theta_{\text{ref}} - \theta_{\text{tach}})$. On integrating the velocity error in a velocity servo, an equivalent to the moving position servo is produced. This shifts the emphasis from phase detection, which is a difficult process, to time integration, which is a relatively easy process. As described in Chapter 3, a digital integrator is easily constructed from up-down counters and a D/A converter.

Both views of the PLL yield useful results. The "position servo" approach shows that it is important to have velocity feedback and use a transconductance amplifier. The velocity servo approach shows that phase detection can be replaced by a pure integration in the forward path. These ideas were used in the formulation of the PLL block diagram.

Permanent Magnet DC Motor Transfer Functions

STATIC EQUATIONS

The permanent magnet DC (PMDC) motor is modeled by the equivalent circuit of Fig. C-1. The motor develops a back EMF ($= nK_e$) proportional to its rotational velocity:

$$\text{back EMF} = E = nK_e \tag{C-1}$$

where n = speed (1000 rpm)
  K_e = back EMF (volts/1000 rpm)
  E = back EMF (volts)

The torque output of the motor is proportional to the current in the winding

$$T = K_T I \tag{C-2}$$

where $T = T_L + T_f + T_D$ = total torque output of the motor (oz-in.)
  T_f = friction torque (oz-in.)
  T_L = load torque (oz-in.)
  T_D = damping torque (oz-in.)
  K_T = torque constant (oz-in./A)
  I = current (A)

The power delivered to the load can be expressed in terms of the voltage and current conditions at the back EMF generator. Note that these differ from the available motor terminals. Since current I flows into the generator, which has a back EMF E, the power into the back EMF generator is

$$P_{\text{in}} = EI \tag{C-3}$$

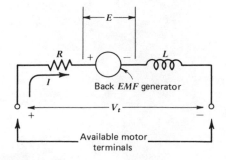

Figure C-1 Equivalent circuit of the permanent magnet DC motor. R = internal winding resistance (Ω); L = winding inductance (H).

162

where P_{in} = power into the back EMF generator (W). Substituting eqs. C-1 and C-2 into eq. C-3 yields

$$P_{in} = nK_e \times \frac{T}{K_T} \tag{C-4}$$

When viewed at its mechanical output (i.e., the motor shaft), the total motor output power is

$$P_{out} = T\omega \tag{C-5}$$

where T = torque in N-m
ω = shaft speed in rad/sec
P = power in watts

Converting torque units to oz-in (instead of N-m), and shaft speed to units of 1000 rpm (instead of rad/sec), yields

$$P_{out} = 0.74Tn \tag{C-6}$$

where T = in oz-in.
n = in 1000 rpm

By the principle of conservation of energy, motor output power P_{out} must equal motor input power P_{in}. (The losses in the motor are accounted by for the T_f and T_D components.)

$$P_{out} = P_{in} \tag{C-7}$$

Substituting eqs. C-4 and C-6 into eq. C-7 yields

$$\boxed{K_e = 0.74 \, K_T} \tag{C-8}$$

This shows that K_e and K_T are not independent, but are related as in eq. C-8. When the motor is viewed from its input terminals, the steady state equation describing the motor is

$$V_t = E + IR \tag{C-9}$$

Substituting eq. C-1 yields

$$V_t = nK_e + IR \tag{C-10}$$

DYNAMIC EQUATIONS

The dynamic equations are based on Newtons law, $F = ma$. This is rewritten in rotational form and with a term to account for viscous damping:

$$T = J\alpha + K_D\omega \tag{C-11}$$

where $T =$ dynamic torque (oz-in.)
        $\alpha =$ angular acceleration (rad/sec^2)
        $\omega =$ angular speed (rad/sec)
        $K_D =$ damping constant (oz-in./rad/sec)
        $J =$ inertia (oz-in./sec^2)

Note that T is the *dynamic* torque. Any steady state torque produced by the motor is ignored, since it does not influence dynamic performance. Note also that there are two components of dynamic torque. One component accelerates the motor $(= J\alpha)$ and the other overcomes damping $(= K_D\omega)$.

Rewriting eq. C-11 using Laplace notation, we have

$$T = Js\omega + K_D\omega \tag{C-12}$$

where $s =$ Laplace operator. Substituting eq. C-2 into eq. C-12 and solving for the ratio ω/I yields

$$\frac{\omega}{I} = \frac{K_T}{Js + K_D} \tag{C-13}$$

This equation is the transfer function between motor speed ω and applied current I. It has only a single pole due to the inertia of the system.

The better known way of viewing the PMDC motor is in terms of speed versus applied voltage (as compared to applied current). From the model of Fig. C-1, the equation governing the motor is written as

$$V_t = nK_e + I(sL + R) \tag{C-14}$$

It is convenient to have the speed in rad/sec. This is easily handled by defining a back EMF constant K_e such that

$$\omega K_e' = nK_e \tag{C-15}$$

Substituting eqs. C-15 and C-13 into eq. C-14, we have

$$V_t = \omega K_e' + \left(\frac{Js + K_D}{K_T}(\omega)\right)(sL + R) \tag{C-16}$$

Solving for the ratio ω/V_t:

$$\frac{\omega}{V_t} = \frac{1/K_e'}{\dfrac{JL}{K_e'K_T}s^2 + s\left[\dfrac{K_D L + JR}{K_e'K_T}\right] + \left[\dfrac{K_D R}{K_e'K_T} + 1\right]} \tag{C-17}$$

Equation C-17 may be rewritten as

$$\frac{\omega}{V_t} = \frac{1/K_e'}{(s\tau_1 + 1)(s\tau_2 + 1)} \tag{C-18}$$

with

$$\tau_1 = \frac{1}{a + b} \tag{C-19}$$

$$\tau_2 = \frac{1}{a - b} \tag{C-20}$$

where

$$a = -\frac{LK_D + JR}{2JL}$$

$$b = \sqrt{\left(\frac{LK_D - JR}{2JL}\right)^2 - \frac{K_e'K_T}{JL}}$$

The roots are usually real, but that depends on the relative values inside the radical sign.

It is often advantageous to use the PMDC motor as a current operated device, primarily because the resulting transfer function (eq. C-13) has only one pole.

D

Inertia
Independent Design

It is often required to design a PLL that is insensitive to changes in inertial load. The servo must be stable regardless of the inertial load being driven, with no need to adjust any parameters. When the system specifications are not stringent a system can usually be designed that is independent of inertial load. The technique is based on the fact that when the inertia changes, only ω_J moves; all other factors that influence the crossover frequency of the Bode plot remain the same. Consider the Bode plot of Fig. D-1. Note that only ω_J has changed, yet both systems shown as solid lines are most likely to be stable. In the high inertia case, the crossover frequency is shifted to a relatively low value, but the slope at crossover is -20 dB/decade, indicating a likelihood of stability. In the low inertia case, the slope at crossover is also at -20 dB/decade, implying stability.

The two cases shown as dotted lines in Fig. D-1 indicate a probable instability for the value of ω_J used. The acceptable range of ω_J (and therefore of J) is clearly between ω_{JH} and ω_{JL}. In the two stable cases it is clear that

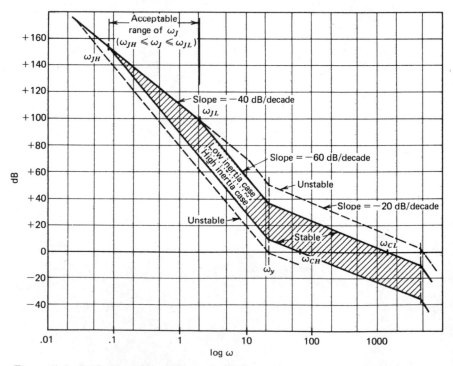

Figure D-1 Bode plots for various values of ω_J. All other significant frequencies (ω_y, ω_M, and ω_f) are held constant. ω_C follows from the choice of ω_J.

168

any inertial load between the two extremes represented by ω_{JH} and ω_{JL} will provide stable operation. Furthermore, the phase margin and system bandwidth will vary as a function of inertial load. As mentioned earlier, if the servo is not too stringently specified, an inertia independent system can be designed. One such design procedure and associated TI59 program is presented below.

INERTIA INDEPENDENT
PLL DESIGN PROCEDURE

The following parameters are assumed to be known:

$$J_H = \text{highest inertia to be used}$$
$$J_L = \text{lowest inertia to be used}$$
$$K_D = \text{damping constant of the motor}$$
$$N = \text{disc density}$$
$$\text{LRPM} = \text{minimum rpm}$$

Procedure:

1 Calculate ω_{JH} and ω_{JL} and mark their location on a sketch of the Bode plot (Fig. D-2). (Plotting paper need not be used.)

$$\omega_{JH} = \frac{K_D}{J_H} \quad (\omega_{JH} = \text{break frequency of the high inertia case})$$

$$\omega_{JL} = \frac{K_D}{J_L} \quad (\omega_{JL} = \text{break frequency of the low inertia case})$$

2 Calculate $\omega_{\text{tach min}}$ from the LRPM specification and mark its location on the sketch.

$$\omega_{\text{tach min}} = \frac{\pi}{30} N \,\text{LRPM} \quad \begin{matrix}(\omega_{\text{tach min}} = \text{minimum radian frequency} \\ \text{from tachometer, at LRPM})\end{matrix}$$

3 Select ω_{CL} to be an order of magnitude less than $\omega_{\text{tach min}}$. This avoids sampled data delay.

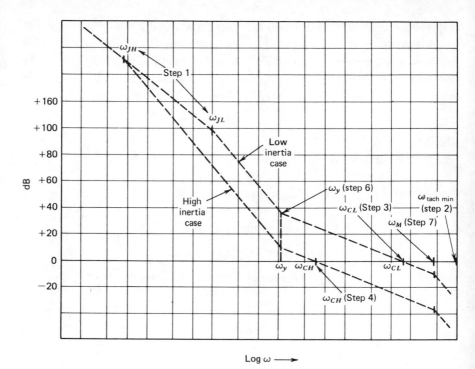

Figure D-2 Significant frequencies and the step of the procedure at which they are determined.

$$\omega_{CL} = \frac{\omega_{\text{tach min}}}{10} \qquad (\omega_{CL} = \text{crossover frequency of the low inertia system})$$

4 Select ω_{CH} according to

$$\omega_{CH} = \omega_{CL}\,\frac{\omega_{JH}}{\omega_{JL}} \qquad (\omega_{CH} = \text{crossover frequency of the high inertia system})$$

This selection of ω_{CH} is based on the fact that the crossover frequencies of the asymptotic approximation will have the same ratio as the inertial break frequencies ω_{JH} and ω_{JL}.

5 Is $\omega_{JL} \leqslant \omega_{CH}$? If yes, a design is possible. If no, a design is not possible.

6 Is $\omega_{JL} \leqslant \omega_{CH}/3$? If yes, select ω_y (the cut-in frequency of the numerator term of the transfer function; see eq. 5-7) as $\omega_{CH}/3$. If no, select ω_y as ω_{JL}.

7 Select ω_M (= ω_f) as 3 ω_{CL}. This is the approximate geometric mean frequency between $\omega_{tach\ min}$ and ω_{CL}. It represents a compromise between good carrier filtering and high phase margin.

8 The gain K can be calculated from eq. 5-23:

$$K = \frac{\omega_c^2 \sqrt{(\omega_c/\omega_J)^2 + 1}\ \sqrt{(\omega_c/\omega_M)^2 + 1}\sqrt{(\omega_c/\omega_f)^2 + 1}}{\sqrt{(2\omega_y\omega_c)^2 + (\omega_y^2 - \omega_c^2)^2}}$$

Setting $\omega_f = \omega_M$, substituting, and simplifying, we find

$$K = \frac{\sqrt{(\omega_c/\omega_J)^2 + 1}\ [(\omega_c/\omega_M)^2 + 1]}{[(\omega_y/\omega_c)^2 + 1]}$$

At this point, the Bode plot of the system is fully defined. Values for ω_{JL}, ω_{CL}, ω_{JH}, ω_{CH}, ω_y, ω_M, and K have been selected. The design can be checked by loading the pertinent parameters into program "ANGL, MAGN, CLR" and obtaining the Nyquist plot and the closed loop response. The reader is cautioned to load the appropriate parameters together, that is, ω_{CH} and ω_{JH} must be used together. Using ω_{CH} and ω_{JL} would produce meaningless results. A flowchart is shown in Fig. D-3 for easy reference to the procedure. The TI59 program "ALL J/BODE" calculates all the important frequencies of the Bode plot from the input data. "ALL J/BODE" is listed on pages 174 and 175.

With the Bode plot defined, it is necessary to determine the parameters G_1, k_i, and k_ρ. The procedure is as follows:

1 Use eq. 5-26 to find D:

$$D = \frac{KK_D}{A_1 K_T N}$$

Substitute the known values for K, K_D, K_T, and N. Select an arbitrary value for A_1 such that $10 \leqslant A_1 \leqslant 1000$. This procedure is identical to the one used in Chapter 6.

2 Use eq. 5-32 to determine k_ρ. (Note that $\omega_c/\eta = \omega_y$.)

$$k_\rho = \omega_y^2 D$$

3   Use eq. 5-33 to determine k_i. (Note that $\omega_c \eta = \omega_M$.)

$$k_i = 2\omega_y D - k_\rho \left(\frac{1}{\omega_y} + \frac{1}{\omega_M} \right)$$

4   Determine G_1 from eq. 5-34:

$$G_1 = \frac{D - k_i/\omega_M - k_\rho/\omega_M \omega_y}{k_m}$$

Note that the product $\omega_M \omega_y$ is substituted for $\omega_c{}^2$ in the denominator of the k_ρ term. This is mathematically correct and represents an "average" crossover frequency. The flowchart for this procedure is shown in Fig. D-4. A TI59 program "ALLJ/GAINS" is provided to calculate k_i, k_ρ, and G_1 from the Bode plot parameters and other system constants (see pages 177 and 178.

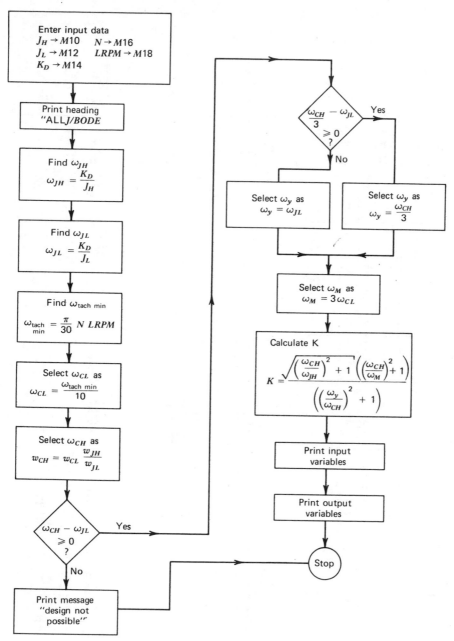

Figure D-3 "ALL J/BODE" flowchart.

ALL J/BODE

<table>
<tr><td>Initialize</td><td>000</td><td>69</td><td>OP</td></tr>
<tr><td></td><td>001</td><td>00</td><td>00</td></tr>
<tr><td></td><td>002</td><td>00</td><td>0</td></tr>
<tr><td></td><td>003</td><td>32</td><td>X:T</td></tr>
<tr><td>Print "ALL J /BODE"</td><td>004</td><td>43</td><td>RCL</td></tr>
<tr><td></td><td>005</td><td>59</td><td>59</td></tr>
<tr><td></td><td>006</td><td>69</td><td>OP</td></tr>
<tr><td></td><td>007</td><td>01</td><td>01</td></tr>
<tr><td></td><td>008</td><td>43</td><td>RCL</td></tr>
<tr><td></td><td>009</td><td>58</td><td>58</td></tr>
<tr><td></td><td>010</td><td>69</td><td>OP</td></tr>
<tr><td></td><td>011</td><td>02</td><td>02</td></tr>
<tr><td></td><td>012</td><td>69</td><td>OP</td></tr>
<tr><td></td><td>013</td><td>05</td><td>05</td></tr>
<tr><td></td><td>014</td><td>69</td><td>OP</td></tr>
<tr><td></td><td>015</td><td>00</td><td>00</td></tr>
<tr><td>ω_{JH}</td><td>016</td><td>43</td><td>RCL</td></tr>
<tr><td></td><td>017</td><td>14</td><td>14</td></tr>
<tr><td></td><td>018</td><td>65</td><td>×</td></tr>
<tr><td></td><td>019</td><td>43</td><td>RCL</td></tr>
<tr><td></td><td>020</td><td>57</td><td>57</td></tr>
<tr><td></td><td>021</td><td>55</td><td>÷</td></tr>
<tr><td></td><td>022</td><td>43</td><td>RCL</td></tr>
<tr><td></td><td>023</td><td>10</td><td>10</td></tr>
<tr><td></td><td>024</td><td>95</td><td>=</td></tr>
<tr><td></td><td>025</td><td>42</td><td>STO</td></tr>
<tr><td></td><td>026</td><td>20</td><td>20</td></tr>
<tr><td>ω_{JL}</td><td>027</td><td>43</td><td>RCL</td></tr>
<tr><td></td><td>028</td><td>14</td><td>14</td></tr>
<tr><td></td><td>029</td><td>65</td><td>×</td></tr>
<tr><td></td><td>030</td><td>43</td><td>RCL</td></tr>
<tr><td></td><td>031</td><td>57</td><td>57</td></tr>
<tr><td></td><td>032</td><td>55</td><td>÷</td></tr>
<tr><td></td><td>033</td><td>43</td><td>RCL</td></tr>
<tr><td></td><td>034</td><td>12</td><td>12</td></tr>
<tr><td></td><td>035</td><td>95</td><td>=</td></tr>
<tr><td></td><td>036</td><td>42</td><td>STO</td></tr>
<tr><td></td><td>037</td><td>22</td><td>22</td></tr>
<tr><td>ω_{CL}</td><td>038</td><td>89</td><td>π</td></tr>
<tr><td></td><td>039</td><td>55</td><td>÷</td></tr>
<tr><td></td><td>040</td><td>03</td><td>3</td></tr>
<tr><td></td><td>041</td><td>00</td><td>0</td></tr>
<tr><td></td><td>042</td><td>00</td><td>0</td></tr>
<tr><td></td><td>043</td><td>65</td><td>×</td></tr>
<tr><td></td><td>044</td><td>43</td><td>RCL</td></tr>
<tr><td></td><td>045</td><td>16</td><td>16</td></tr>
<tr><td></td><td>046</td><td>65</td><td>×</td></tr>
<tr><td></td><td>047</td><td>43</td><td>RCL</td></tr>
<tr><td></td><td>048</td><td>18</td><td>18</td></tr>
<tr><td></td><td>049</td><td>95</td><td>=</td></tr>
<tr><td></td><td>050</td><td>42</td><td>STO</td></tr>
<tr><td></td><td>051</td><td>24</td><td>24</td></tr>
<tr><td></td><td>052</td><td>65</td><td>×</td></tr>
<tr><td></td><td>053</td><td>43</td><td>RCL</td></tr>
<tr><td></td><td>054</td><td>20</td><td>20</td></tr>
</table>

<table>
<tr><td>ω_{CH}</td><td>055</td><td>55</td><td>÷</td></tr>
<tr><td></td><td>056</td><td>43</td><td>RCL</td></tr>
<tr><td></td><td>057</td><td>22</td><td>22</td></tr>
<tr><td></td><td>058</td><td>95</td><td>=</td></tr>
<tr><td></td><td>059</td><td>42</td><td>STO</td></tr>
<tr><td></td><td>060</td><td>26</td><td>26</td></tr>
<tr><td>Test $\omega_{JL} \leqslant \omega_{CH}$</td><td>061</td><td>75</td><td>-</td></tr>
<tr><td></td><td>062</td><td>43</td><td>RCL</td></tr>
<tr><td></td><td>063</td><td>22</td><td>22</td></tr>
<tr><td></td><td>064</td><td>95</td><td>=</td></tr>
<tr><td></td><td>065</td><td>77</td><td>GE</td></tr>
<tr><td></td><td>066</td><td>33</td><td>X²</td></tr>
<tr><td>Print "DESIGN NOT POSSIBLE" if not true</td><td>067</td><td>43</td><td>RCL</td></tr>
<tr><td></td><td>068</td><td>56</td><td>56</td></tr>
<tr><td></td><td>069</td><td>69</td><td>OP</td></tr>
<tr><td></td><td>070</td><td>01</td><td>01</td></tr>
<tr><td></td><td>071</td><td>43</td><td>RCL</td></tr>
<tr><td></td><td>072</td><td>55</td><td>55</td></tr>
<tr><td></td><td>073</td><td>69</td><td>OP</td></tr>
<tr><td></td><td>074</td><td>02</td><td>02</td></tr>
<tr><td></td><td>075</td><td>43</td><td>RCL</td></tr>
<tr><td></td><td>076</td><td>54</td><td>54</td></tr>
<tr><td></td><td>077</td><td>69</td><td>OP</td></tr>
<tr><td></td><td>078</td><td>03</td><td>03</td></tr>
<tr><td></td><td>079</td><td>43</td><td>RCL</td></tr>
<tr><td></td><td>080</td><td>53</td><td>53</td></tr>
<tr><td></td><td>081</td><td>69</td><td>OP</td></tr>
<tr><td></td><td>082</td><td>04</td><td>04</td></tr>
<tr><td></td><td>083</td><td>69</td><td>OP</td></tr>
<tr><td></td><td>084</td><td>05</td><td>05</td></tr>
<tr><td></td><td>085</td><td>69</td><td>OP</td></tr>
<tr><td></td><td>086</td><td>00</td><td>00</td></tr>
<tr><td></td><td>087</td><td>91</td><td>R/S</td></tr>
<tr><td></td><td>088</td><td>81</td><td>RST</td></tr>
<tr><td>ω_y</td><td>089</td><td>76</td><td>LBL</td></tr>
<tr><td></td><td>090</td><td>33</td><td>X²</td></tr>
<tr><td></td><td>091</td><td>43</td><td>RCL</td></tr>
<tr><td></td><td>092</td><td>26</td><td>26</td></tr>
<tr><td></td><td>093</td><td>55</td><td>÷</td></tr>
<tr><td></td><td>094</td><td>03</td><td>3</td></tr>
<tr><td></td><td>095</td><td>75</td><td>-</td></tr>
<tr><td></td><td>096</td><td>43</td><td>RCL</td></tr>
<tr><td></td><td>097</td><td>22</td><td>22</td></tr>
<tr><td></td><td>098</td><td>95</td><td>=</td></tr>
<tr><td></td><td>099</td><td>77</td><td>GE</td></tr>
<tr><td></td><td>100</td><td>34</td><td>ΓX</td></tr>
<tr><td></td><td>101</td><td>43</td><td>RCL</td></tr>
<tr><td></td><td>102</td><td>22</td><td>22</td></tr>
<tr><td></td><td>103</td><td>61</td><td>GTO</td></tr>
<tr><td></td><td>104</td><td>35</td><td>1/X</td></tr>
<tr><td></td><td>105</td><td>76</td><td>LBL</td></tr>
<tr><td></td><td>106</td><td>34</td><td>ΓX</td></tr>
<tr><td></td><td>107</td><td>43</td><td>RCL</td></tr>
<tr><td></td><td>108</td><td>26</td><td>26</td></tr>
<tr><td></td><td>109</td><td>55</td><td>÷</td></tr>
</table>

<table>
<tr><td></td><td>110</td><td>03</td><td>3</td></tr>
<tr><td></td><td>111</td><td>95</td><td>=</td></tr>
<tr><td></td><td>112</td><td>76</td><td>LBL</td></tr>
<tr><td></td><td>113</td><td>35</td><td>1/X</td></tr>
<tr><td></td><td>114</td><td>42</td><td>STO</td></tr>
<tr><td></td><td>115</td><td>28</td><td>28</td></tr>
<tr><td></td><td>116</td><td>43</td><td>RCL</td></tr>
<tr><td></td><td>117</td><td>24</td><td>24</td></tr>
<tr><td></td><td>118</td><td>65</td><td>×</td></tr>
<tr><td></td><td>119</td><td>03</td><td>3</td></tr>
<tr><td></td><td>120</td><td>95</td><td>=</td></tr>
<tr><td></td><td>121</td><td>42</td><td>STO</td></tr>
<tr><td></td><td>122</td><td>30</td><td>30</td></tr>
<tr><td></td><td>123</td><td>53</td><td>(</td></tr>
<tr><td></td><td>124</td><td>53</td><td>(</td></tr>
<tr><td></td><td>125</td><td>43</td><td>RCL</td></tr>
<tr><td></td><td>126</td><td>26</td><td>26</td></tr>
<tr><td></td><td>127</td><td>55</td><td>÷</td></tr>
<tr><td></td><td>128</td><td>43</td><td>RCL</td></tr>
<tr><td></td><td>129</td><td>20</td><td>20</td></tr>
<tr><td></td><td>130</td><td>54</td><td>)</td></tr>
<tr><td></td><td>131</td><td>33</td><td>X²</td></tr>
<tr><td></td><td>132</td><td>85</td><td>+</td></tr>
<tr><td></td><td>133</td><td>01</td><td>1</td></tr>
<tr><td></td><td>134</td><td>54</td><td>)</td></tr>
<tr><td></td><td>135</td><td>34</td><td>ΓX</td></tr>
<tr><td></td><td>136</td><td>65</td><td>×</td></tr>
<tr><td></td><td>137</td><td>53</td><td>(</td></tr>
<tr><td></td><td>138</td><td>53</td><td>(</td></tr>
<tr><td></td><td>139</td><td>43</td><td>RCL</td></tr>
<tr><td></td><td>140</td><td>26</td><td>26</td></tr>
<tr><td></td><td>141</td><td>55</td><td>÷</td></tr>
<tr><td></td><td>142</td><td>43</td><td>RCL</td></tr>
<tr><td></td><td>143</td><td>30</td><td>30</td></tr>
<tr><td></td><td>144</td><td>54</td><td>)</td></tr>
<tr><td></td><td>145</td><td>33</td><td>X²</td></tr>
<tr><td></td><td>146</td><td>85</td><td>+</td></tr>
<tr><td></td><td>147</td><td>01</td><td>1</td></tr>
<tr><td></td><td>148</td><td>54</td><td>)</td></tr>
<tr><td></td><td>149</td><td>55</td><td>÷</td></tr>
<tr><td></td><td>150</td><td>53</td><td>(</td></tr>
<tr><td></td><td>151</td><td>53</td><td>(</td></tr>
<tr><td></td><td>152</td><td>43</td><td>RCL</td></tr>
<tr><td></td><td>153</td><td>28</td><td>28</td></tr>
<tr><td>K</td><td>154</td><td>55</td><td>÷</td></tr>
<tr><td></td><td>155</td><td>43</td><td>RCL</td></tr>
<tr><td></td><td>156</td><td>26</td><td>26</td></tr>
<tr><td></td><td>157</td><td>54</td><td>)</td></tr>
<tr><td></td><td>158</td><td>33</td><td>X²</td></tr>
<tr><td></td><td>159</td><td>85</td><td>+</td></tr>
<tr><td></td><td>160</td><td>01</td><td>1</td></tr>
<tr><td></td><td>161</td><td>54</td><td>)</td></tr>
<tr><td></td><td>162</td><td>95</td><td>=</td></tr>
<tr><td></td><td>163</td><td>42</td><td>STO</td></tr>
<tr><td></td><td>164</td><td>32</td><td>32</td></tr>
</table>

ALL J/BODE (cont.)

165	43	RCL
166	52	52
167	69	□P
168	01	01
169	69	□P
170	05	05
171	69	□P
172	00	00
173	01	1
174	09	9
175	42	STO
176	05	05
177	01	1
178	00	0
179	32	X:T
180	76	LBL
181	39	COS
182	73	RC+
183	05	05
184	69	□P
185	04	04
186	69	□P
187	35	35
188	73	RC+
189	05	05
190	69	□P
191	06	06
192	69	□P
193	35	35
194	43	RCL
195	05	05
196	77	GE
197	39	COS
198	69	□P
199	00	00
200	43	RCL
201	51	51
202	69	□P

- Print "INPUT" (lines 165–172)
- Input variable print loop (lines 180–197)
- Print "OUTPUT" (lines 198–202)

203	01	01
204	43	RCL
205	50	50
206	69	□P
207	02	02
208	69	□P
209	05	05
210	69	□P
211	00	00
212	03	3
213	03	3
214	42	STO
215	05	05
216	02	2
217	00	0
218	32	X:T
219	76	LBL
220	38	SIN
221	73	RC+
222	05	05
223	69	□P
224	04	04
225	69	□P
226	35	35
227	73	RC+
228	05	05
229	69	□P
230	06	06
231	69	□P
232	35	35
233	43	RCL
234	05	05
235	77	GE
236	38	SIN
237	51	R/S
238	81	RST

- Output variable print loop (lines 219–236)
- Stop (lines 237–238)

Value	Reg
0.	00
0.	01
0.	02
0.	03
0.	04
19.	05
0.	06
0.	07
0.	08
0.	09
$0.522\,J_H$	10
25230000. "JH"	11
$0.022\,J_L$	12
25270000. "JL"	13
$4.7\,K_D$	14
26160000. "KD"	15
5000. N	16
31000000. "N"	17
30. LRPM	18
27353330. "LRPM"	19
.0859802566 W_{JH}	20
43252300. "WJH"	21
2.040076998 ω_{JL}	22
43252700. "WJL"	23
1570.796327 ω_{CL}	24
43152700. "WCL"	25
66.20214404 ω_{CH}	26
43152300. "WCH"	27
22.06738135 ω_y	28
43450000. "WY"	29
4712.38898 ω_M	30
43300000. "WM"	31
693.1095743 K	32
26000000. "K"	33
0.	34
0.	35
0.	36
0.	37
0.	38
0.	39
0.	40
0.	41
0.	42
0.	43
0.	44
0.	45
0.	46
0.	47
0.	48
0.	49
3700000000. T	50
3241373341. OUTPU	51
2431334137. INPUT	52
2414271700. IBLE−	53
33323636. −POSS	54
3100313237. N−NOT	55
1617362422. DESIG	56
.0095492966 $3/100\pi$	57
6314321617. /BODE	58
1327270025. ALL−J	59

175

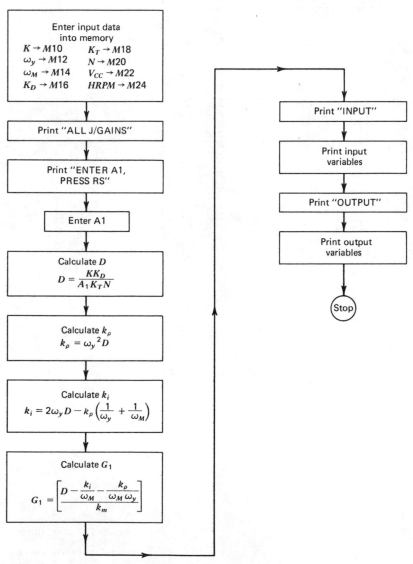

Figure D-4 "ALL J/GAINS" flowchart.

ALLJ/GAINS

```
Print "ALLJ/GAINS"
000  69  □P
001  00  00
002  43  RCL
003  39  39
004  69  □P
005  01  01
006  43  RCL
007  44  44
008  69  □P
009  02  02
010  43  RCL
011  45  45
012  69  □P
013  03  03
014  69  □P
015  05  05
Print "ENTER A," PRESS R/S:
016  69  □P
017  00  00
018  43  RCL
019  40  40
020  69  □P
021  01  01
022  43  RCL
023  41  41
024  69  □P
025  02  02
026  43  RCL
027  42  42
028  69  □P
029  03  03
030  43  RCL
031  43  43
032  69  □P
033  69  □P
034  69  □P
035  05  05
Wait
036  91  R/S
037  42  STO
038  26  26
039  35  1/X
040  65  ×

041  43  RCL
042  10  10
043  65  ×
044  43  RCL
045  16  16
046  65  ×
047  03  3
048  55  ÷
049  89  π
050  55  ÷
051  01  1
052  00  0
053  00  0
054  55  ÷
055  43  RCL
056  18  18
057  55  ÷
058  43  RCL
059  20  20
060  95  =
061  42  STO
062  28  28
063  65  ×
064  43  RCL
065  12  12
066  33  X2
067  95  =
068  42  STO
069  30  30
070  65  ×
071  53  (
072  43  RCL
073  14  14
074  35  1/X
075  85  +
076  43  RCL
077  12  12
078  35  1/X
079  54  )
080  94  +/-

081  85  +
082  02  2
083  65  ×
084  43  RCL
085  12  12
086  65  ×
087  43  RCL
088  28  28
089  95  =
090  42  STO
091  32  32
092  55  ÷
093  43  RCL
094  14  14
095  94  +/-
096  75  -
097  43  RCL
098  30  30
099  55  ÷
100  43  RCL
101  14  14
102  55  ÷
103  43  RCL
104  12  12
105  85  +
106  43  RCL
107  28  28
108  95  =
109  55  ÷
110  53  (
111  00  0
112  55  ÷
113  89  π
114  65  ×
115  43  RCL
116  22  22
117  55  ÷
118  43  RCL
119  20  20
120  55  ÷
121

G,
(cont.)
122  43  RCL
123  24  24
124  54  )
125  95  =
126  42  STO
127  34  34
Print "INPUT"
128  69  □P
129  00  00
130  43  RCL
131  36  36
132  69  □P
133  01  01
134  69  □P
135  05  05
136  02  2
137  07  7
138  42  STO
139  08  08
140  09  9
141  42  STO
142  09  09
143  76  LBL
144  30  TAN
145  73  RC+
146  08  08
147  69  □P
148  04  04
Print loop
149  69  □P
150  38  38
151  73  RC+
152  08  08
153  69  □P
154  06  06
155  38  38
156  97  DSZ
157  22  22
158  09  09
159  30  TAN
Print "OUTPUT"
160  69  □P
161  00  00
162  43  RCL
```

177

	163	37	37	0.	00
	164	69	OP	0.	01
	165	01	01	0.	02
	166	43	RCL	0.	03
	167	38	38	0.	04
	168	69	OP	0.	05
	169	02	02	0.	06
	170	69	OP	0.	07
	171	05	05	29.	08
	172	03	3	0.	09
	173	05	5	693. K	10
	174	42	STO	26000000.	"K"11
	175	08	08	22. ω_y	12
	176	03	3	43450000.	"Wy"13
Output print loop	177	42	STO	4700. ω_M	14
	178	09	09	43300000.	"WM"15
	179	76	LBL	4.7 K_D	16
	180	39	COS	26160000.	"KD"17
	181	73	RC*	27. K_T	18
	182	08	08	26370000.	"KT"19
	183	69	OP	5000. N	20
	184	04	04	31000000.	"N"21
	185	69	OP	15. V_{CC}	22
	186	38	38	42151500.	"VCC"23
	187	73	RC*	3000. HRPM	24
	188	08	08	23353330.	"HRPM"25
	189	69	OP	25. A_1	26
	190	06	06	13020000.	"A1"27
	191	69	OP	.0000092157 D	28
	192	38	38	0.	29
	193	97	DSZ	.0044604026 k_ρ	30
	194	09	09	26352332.	"KRHO"31
	195	39	COS	.0002017966 k_i	32
End	196	91	R/S	26240000.	"KI"33
	197	81	RST	0.956053145 G_1	34
				22020000.	"G1"35
				2431334137.	INPUT 36
				3241373341.	OUTPU 37
				3700000000.	Γ- - - - 38
				1327270025.	ALL-J 39
				1731371735.	ENTER 40
				13025733.	−A1, P 41
				3517363600.	RESS— 42
				3536000000.	RS - - - 43
				6322132431.	/GAIN 44
				3600000000.	S 45

Design Example 3

Redesign the system of Design Example 1 in Chapter 6 so that it can be used with an added inertia of 0.5 oz-in./sec^2 as well as zero inertial load. The pertinent data are repeated here for convenient reference:

$$N = 5000$$

$$\text{LRPM} = 30 \text{ rpm}$$

$$\text{HRPM} = 3000 \text{ rpm}$$

$$V_{cc} = 15 \text{ V}$$

$$K_T = 27 \text{ oz-in/A.}$$

$$K_D = 4.7 \text{ oz-in./1000 rpm}$$

$$J_L = 0.022 \text{ oz-in-sec}^2 \text{ (motor + disc)}$$

$$J_H = 0.522 \text{ oz-in-sec}^2 \text{ (motor + disc + load)}$$

The first step in this design is to use the program "ALL J/BODE" to design the Bode plot of the system. "ALL J/BODE" is loaded into the TI59, and data one put into memory as shown below:

$$0.522 \rightarrow \text{M10 } (J_H, \text{ oz-in-sec}^2)$$

$$0.022 \rightarrow \text{M12 } (J_L, \text{ oz-in-sec}^2)$$

$$4.7 \rightarrow \text{M14 } (K_D, \text{ oz-in./1000 rpm})$$

$$5000 \rightarrow \text{M16 } (N)$$

$$30 \rightarrow \text{M18 (LRPM, in rpm)}$$

Then press "RST" and "RS," and the following printout appears:

```
ALL J/BODE
INPUT
           30.        LRPM
         5000.        N
            4.7       KD
            0.022     JL
            0.522     JH
OUTPUT
      693.1095743     K
     4712.38898       WM
       22.06738135    WY
       66.20214404    WCH
     1570.796327      WCL
        2.040076998   WJL
         .0859802566  WJH
```

The printout provides all Bode plot information. It is prudent to do a Nyquist plot for both the low and high inertia systems. This gives a check on the relative stability of the system at the extremes of inertial load. The program "ANGL, MAGN, CLR" is used. After loading "ANGL, MAGN, CLR," data are put into memory. Note that since the Bode plot changes for the different inertias, two Nyquist plots (for $\omega_J = \omega_{JH}$ and $\omega_J = \omega_{JL}$) are required. The two cases are evaluated as follows:

High Inertia Case	Low Inertia Case
$693 \to$ M02 (gain)	$693 \to$ M02 (gain)
$4700 \to$ M38 (ω_M)	$4700 \to$ M38 (ω_M)
$22 \to$ M36 (ω_y)	$22 \to$ M36 (ω_y)
$0.086 \to$ M34 (ω_{JH})	$2 \to$ M34 (ω_{JL})

ANGL, MAGN, CLR

```
                693.    K
               4700.    WM
                 22.    WY
                0.086   WJ
                 10.    W
         34.80378742    MAGN
        -220.8631703    ANGL
          .1892188522   CLR
                 20.    W
          6.58539887    MAGN
        -185.6938724    ANGL
          1.421520953   CLR
                 50.    W
          1.422560356   MAGN
        -138.6194535    ANGL
          3.572969378   CLR
                100.    W
           .6245424746  MAGN
        -117.2033132    ANGL
         -3.221784595   CLR
                200.    W
           .3010505166  MAGN
        -107.4032563    ANGL
        -10.02023133    CLR
                 30.    W
          3.054812321   MAGN
        -163.0748536    ANGL
          3.180239806   CLR
                 25.    W
          4.229882942   MAGN
        -173.1079816    ANGL
          2.317457045   CLR
```

ANGL, MAGN, CLR

```
                693.    K
               4700.    WM
                 22.    WY
                  2.    WJ
                 10.    W
         793.7019398    MAGN
        -210.0459694    ANGL
           .0094764068  CLR
                 20.    W
         152.3901714    MAGN
        -180.2296496    ANGL
           .0571850731  CLR
                 50.    W
          33.05641334   MAGN
        -136.4273921    ANGL
           .1905135236  CLR
                100.    W
          14.52134499   MAGN
        -116.1068247    ANGL
           .2496568648  CLR
                200.    W
           7.00082542   MAGN
        -106.8549548    ANGL
           .2799615903  CLR
                500.    W
          2.746263975   MAGN
        -106.9545095    ANGL
           .3610776431  CLR
               1000.    W
          1.326613133   MAGN
        -116.4289731    ANGL
           .4711111149  CLR
               2000.    W
           .5868230992  MAGN
        -137.3057624    ANGL
         -1.458087514   CLR
               5000.    W
           .1300375039  MAGN
        -184.0242208    ANGL
        -16.51231609    CLR
```

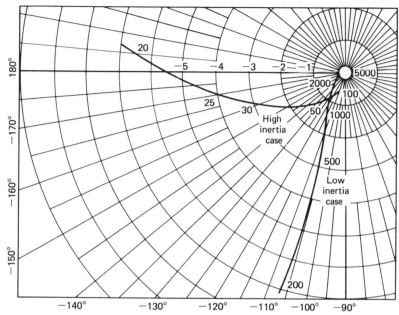

Figure D-5 Nyquist plots for Example 3. Both values of J yield stable servos.

The Nyquist plots for the two extremes of inertial load (Fig. D-5) differ greatly in appearance. Yet a phase margin of better than $50°$ exists for both systems. This is an entirely satisfactory situation.

The last step is to determine gains G_1, k_i, and k_ρ using the program "ALL J/GAINS." After loading, data are entered as

$$693 \rightarrow M10 \ (K)$$

$$22 \rightarrow M12 \ (\omega_y)$$

$$4700 \rightarrow M14 \ (\omega_M)$$

$$4.7 \rightarrow M16 \ (K_D)$$

$$27 \rightarrow M18 \ (K_T)$$

$$5000 \rightarrow M20 \ (N)$$

$$15 \rightarrow M22 \ (V_{cc})$$

$$3000 \rightarrow M24 \ (HRPM)$$

Pressing "RST" and "RS" gives the printouts shown below. Values of $A_1 = 100$ and $A_1 = 25$ were used. It is probably best to use $A_1 \leqslant 25$. This keeps k_i and k_ρ as high as possible.

```
ALL J/GAINS                  ALL J/GAINS
ENTER A1,PRESS RS            ENTER A1,PRESS RS
INPUT                        INPUT
          100.    A1                   25.    A1
         3000.    HRPM               3000.    HRPM
           15.    VCC                  15.    VCC
         5000.    N                  5000.    N
           27.    KT                   27.    KT
            4.7   KD                    4.7   KD
         4700.    WM                 4700.    WM
           22.    WY                   22.    WY
          693.    K                    693.    K
OUTPUT                       OUTPUT
  .2390132862    G1          0.956053145    G1
  .0000504491    KI            .0002017966  KI
  .0011151006    KRHO          .0044604026  KRHO
```

E

The Quadratic Term and the Second Order System

THE QUADRATIC TERM

Quadratic terms often appear in both the numerator and denominator of transfer functions:

$$GH(s) = \frac{\cdots (s^2 + 2\zeta\omega_N s + \omega_N{}^2) \cdots}{\cdots (s^2 + 2\zeta_1\omega_{N_1} s + \omega_{N_1}{}^2) \cdots} \qquad \text{(E-1)}$$

The numerator term may be rewritten in the pure frequency domain as

$$F(j\omega) = (j\omega)^2 + 2\zeta\omega_N(j\omega) + \omega_N{}^2 \qquad \text{(E-2)}$$

Collecting real and imaginary terms:

$$F(j\omega) = 2\zeta\omega_N(j\omega) + (\omega_N{}^2 - \omega^2) \qquad \text{(E-3)}$$

Factoring out $\omega_N{}^2$, we have

$$F(j\omega) = \omega_N{}^2 \left[2\zeta j \frac{\omega}{\omega_N} + \left(1 - \left(\frac{\omega}{\omega_N} \right)^2 \right) \right] \qquad \text{(E-4)}$$

Substituting the variable x for ω/ω_N:

$$F(x) = \omega_N{}^2 [2\zeta jx + (1 - x^2)] \qquad \text{(E-5)}$$

The magnitude and angle of this function can be written as

$$|F(x)| = \omega_N{}^2\sqrt{(2\zeta x)^2 + (1 - x^2)^2} = \omega_N{}^2\sqrt{x^4 + x^2(4\zeta^2 - 2) + 1} \qquad \text{(E-6)}$$

$$\measuredangle F(x) = \tan^{-1}\frac{2\zeta x}{1 - x^2} \qquad \text{(E-7)}$$

In dB, the magnitude is

$$|F(x)|_{\text{dB}} = 20\log\omega_N{}^2 + 20\log\sqrt{x^4 + x^2(4\zeta^2 - 2) + 1} \qquad \text{(E-8)}$$

The first term of eq. E-8 is a constant that describes the "DC gain." The second term is the frequency dependent portion and is separated as $G(x)$:

184

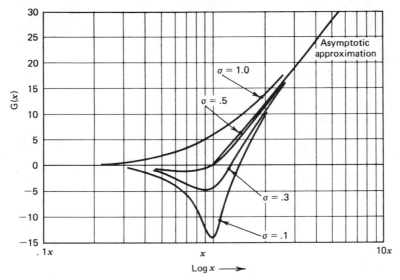

Figure E-1   Magnitude versus frequency.

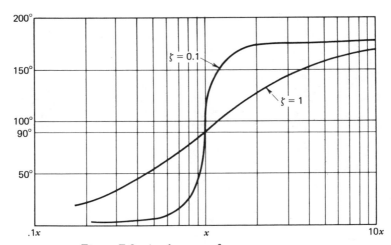

Figure E-2   Angle versus frequency.

$$|G(x)|_{dB} = 20\log\sqrt{x^4 + x^2(4\zeta^2 - 2) + 1} \qquad \text{(E-9)}$$

A T159 program is provided on pages 186, 187 and 188 to plot the magnitude and angle (Figs. E-1 and E-2 and eqs. E-9 and E-7) as a function of normalized frequency (x). The resolution of the points was increased for $x \leqslant 3$. Note that as ζ decreases, the peaking becomes more "dramatic" and the rate of change of phase is faster.

Print "ZETA" and its value

000	43	RCL
001	15	15
002	69	OP
003	04	04
004	43	RCL
005	11	11
006	69	OP
007	06	06

Loop for x ≥ 3

008	03	3
009	32	X:T
010	01	1
011	00	0
012	42	STO
013	00	00
014	76	LBL
015	39	COS
016	43	RCL
017	00	00
018	67	EQ
019	30	TAN
020	42	STO
021	10	10
022	71	SBR
023	34	ΓX
024	71	SBR
025	35	1/X
026	71	SBR
027	99	PRT
028	69	OP
029	30	30
030	61	GTO
031	39	COS

Loop for x < 3

032	76	LBL
033	30	TAN
034	03	3
035	00	0
036	42	STO
037	00	00
038	76	LBL
039	38	SIN
040	43	RCL
041	00	00
042	55	÷
043	01	1
044	00	0
045	95	=
046	42	STO
047	10	10
048	71	SBR
049	34	ΓX
050	71	SBR
051	35	1/X
052	71	SBR
053	99	PRT
054	97	DSZ
055	00	00
056	38	SIN

End

057	91	R/S

Find dB

058	81	RST
059	76	LBL
060	34	ΓX
061	53	(
062	53	(
063	53	(
064	43	RCL
065	10	10
066	45	YX
067	04	4
068	54	)
069	85	+
070	43	RCL
071	10	10
072	33	X²
073	65	×
074	53	(
075	43	RCL
076	11	11
077	33	X²
078	65	×
079	04	4
080	75	-
081	02	2
082	54	)
083	85	+
084	01	1
085	54	)
086	34	ΓX
087	28	LOG
088	65	×
089	02	2
090	00	0
091	54	)
092	42	STO
093	12	12
094	92	RTN

Find x

095	76	LBL
096	35	1/X
097	53	(
098	43	RCL
099	10	10
100	65	×
101	03	3
102	54	)
103	22	INV
104	67	EQ
105	43	RCL
106	09	9
107	00	0
108	42	STO
109	16	16
110	92	RTN $x = 1$
111	76	LBL
112	43	RCL
113	53	(
114	53	(

115	43	RCL
116	10	10
117	65	×
118	02	2
119	65	×
120	43	RCL
121	11	11
122	55	÷
123	53	(
124	01	1
125	75	-
126	43	RCL
127	10	10
128	33	X²
129	54	)
130	54	)
131	85	+
132	03	3
133	54	)
134	22	INV
135	77	GE
136	42	STO
137	53	(
138	24	CE
139	75	-
140	03	3
141	54	)
142	22	INV
143	30	TAN
144	42	STO
145	16	16
146	92	RTN $x < 1$
147	76	LBL
148	42	STO
149	53	(
150	53	(
151	24	CE
152	75	-
153	03	3
154	54	)
155	22	INV
156	30	TAN
157	85	+
158	01	1
159	08	8
160	00	0
161	54	)
162	42	STO
163	16	16
164	92	RTN $x > 1$

PRINT LOOP

165	76	LBL
166	99	PRT
167	43	RCL
168	13	13
169	69	OP
170	04	04
171	43	RCL

Program to find data points
for $|G(x)|$ dB (eq. E–9) and $\angle G(x)$ (eq E–7)

186

172	10	10	0. 1	ZETA	1. 9	W
173	69	OP	10.	W	8. 423907895	DB
174	06	06	39. 91447598	DB	171. 71629	ANGL
175	43	RCL	178. 8426669	ANGL	1. 8	W
176	14	14	9.	W	7. 115710436	DB
177	69	OP	38. 0639978	DB	170. 8698235	ANGL
178	04	04	178. 7110624	ANGL	1. 7	W
179	43	RCL	8.	W	5. 667555835	DB
180	12	12	35. 98961128	DB	169. 8018992	ANGL
181	69	OP	178. 5451818	ANGL	1. 6	W
182	06	06	7.	W	4. 041492492	DB
183	43	RCL	33. 6285177	DB	168. 4078246	ANGL
184	17	17	178. 3293467	ANGL	1. 5	W
185	69	OP	6.	W	2. 181414682	DB
186	04	04	30. 88646306	DB	166. 5042667	ANGL
187	43	RCL	178. 0363425	ANGL	1. 4	W
188	16	16	5.	W	8. 6858896 -12 DB
189	69	OP	27. 61175813	DB	163. 7397953	ANGL
190	06	06	177. 614056	ANGL	1. 3	W
191	92	RTN	4.	W	-2. 64640667	DB
192	00	0	23. 53416091	DB	159. 3530092	ANGL
193	00	0	176. 9471175	ANGL	1. 2	W
194	00	0	3.	W	-5. 999803645	DB
195	00	0	18. 08616035	DB	151. 3895403	ANGL
196	00	0	175. 7108467	ANGL	1. 1	W
197	00	0	2. 9	W	-10. 33858267	DB
198	00	0	17. 42289047	DB	133. 667780	ANGL

Print loop (brace spanning rows 182–191)

175. 5244352	ANGL	1.	W
2. 8	W	-13. 97940009	DB
16. 73013531	DB	90.	ANGL
175. 3195564	ANGL	0. 9	W

Memory contents

0. 1 ω	10	
0. 1 ζ	11	
-0. 08552402 dB	12	
43000000.	"W"	13
16140000.	"DB"	14
46173713.	"ZETA"	5
1. 157333068 x	16	
13312227.	"ANGL"	17

2. 7	W	-11. 64309429	DB
16. 00490441	DB	43. 4518423	ANGL
175. 0931568	ANGL	0. 8	W
2. 6	W	-8. 091082831	DB
15. 24370155	DB	23. 96248897	ANGL
174. 8414482	ANGL	0. 7	W
2. 5	W	-5. 533075336	DB
14. 44240028	DB	15. 05013649	ANGL
174. 559668	ANGL	0. 6	W
2. 4	W	-3. 726341434	DB
13. 59607845	DB	10. 61965528	ANGL
174. 2417398	ANGL	0. 5	W
2. 3	W	-2. 42224509	DB
12. 69879371	DB	7. 594643369	ANGL
173. 8797807	ANGL	0. 4	W
2. 2	W	-1. 475200064	DB
11. 74327341	DB	5. 440332031	ANGL
173. 4633664	ANGL	0. 3	W
2. 1	W	-. 8003329852	DB
10. 72047597	DB	3. 772283609	ANGL
172. 9784055	ANGL	0. 2	W
2.	W	-. 3470420419	DB
9. 618954737	DB	2. 38594403	ANGL
172. 4053566	ANGL	0. 1	W
		-0. 08552402	DB
		1. 157333068	ANGL

187

Left		Right	
1.	ZETA	1.9	W
10.	W	13.27401851	DB
40.08642748	DB	124.4829188	ANGL
168.5788137	ANGL	1.8	W
9.	W	12.54731713	DB
38.27627705	DB	121.8907918	ANGL
167.3196165	ANGL	1.7	W
8.	W	11.79899203	DB
36.25826713	DB	119.0689102	ANGL
165.7499673	ANGL	1.6	W
7.	W	11.02899996	DB
33.97940009	DB	115.9892336	ANGL
163.7397953	ANGL	1.5	W
6.	W	10.23766722	DB
31.36403448	DB	112.6198649	ANGL
161.0753556	ANGL	1.4	W
5.	W	9.425834221	DB
28.29946696	DB	108.9246444	ANGL
157.3801351	ANGL	1.3	W
4.	W	8.5950456	DB
24.60897843	DB	104.8628159	ANGL
151.9275131	ANGL	1.2	W
3.	W	7.747796527	DB
20.	DB	100.3888578	ANGL
143.1301024	ANGL	1.1	W
2.9	W	6.887845474	DB
19.47179247	DB	95.45262199	ANGL
141.9487879	ANGL	1.	W
2.8	W	6.020599913	DB
18.9290453	DB	90.	ANGL
140.6923519	ANGL	0.9	W
2.7	W	5.153571497	DB
18.37109061	DB	83.97442499	ANGL
139.3537263	ANGL	0.8	W
2.6	W	4.296876961	DB
17.79723443	DB	77.31961651	ANGL
137.9249779	ANGL	0.7	W
2.5	W	3.463725368	DB
17.20676013	DB	69.9840404	ANGL
136.397181	ANGL	0.6	W
2.4	W	2.670778167	DB
16.59893392	DB	61.92751306	ANGL
134.7602701	ANGL	0.5	W
2.3	W	1.93820026	DB
15.97301291	DB	53.13010235	ANGL
133.0028686	ANGL	0.4	W
2.2	W	1.289159785	DB
15.32825694	DB	43.60281897	ANGL
131.1120904	ANGL	0.3	W
2.1	W	.7485299588	DB
14.6639453	DB	33.39848847	ANGL
129.0733099	ANGL	0.2	W
2.	W	0.340666786	DB
13.97940009	DB	22.61986495	ANGL
126.8698976	ANGL	0.1	W
		.0864274757	DB
		11.42118627	ANGL

Typical run for $\zeta = 0.1$

THE SECOND ORDER SYSTEM

A second order system is described by the (closed loop) transfer function:

$$G(s) = \frac{\omega_N{}^2}{s^2 + 2\zeta\omega_N + \omega_N{}^2} \qquad \text{(E-10)}$$

The quadratic factor appears in the denominator of $G(s)$. By using $s = j\omega$, angle and magnitude curves can be generated exactly as in the case of the quadratic factor in the numerator. The curves for the denominator case are the mirror image (about the $\log\omega$ axis) of the curves for the numerator case. Instead of a slope of $+40$ dB/decade, the denominator plot exhibits a slope of -40 dB/decade. The simple expedient of relabeling the $+$dB axis as the $-$dB axis (change the sign of the axis) generates the proper curve.

The time domain response of the second order system depends on the driving function. It is helpful to obtain the response of the second order system to a unit step input, since this yields a clear picture of the transient performance. The time domain response $g(t)$ of the second order system to a unit step, for the *underdamped case only*, is

$$g(t) = 1 - \frac{e^{-\zeta\omega_N t}}{\sqrt{1-\zeta^2}} \sin(\omega_N\sqrt{1-\zeta^2}\,t - \psi) \qquad \text{(E-11)}$$

$$\psi = \tan^{-1}\frac{\sqrt{1-\zeta^2}}{-\zeta} \qquad \text{(E-12)}$$

where ζ = damping factor and $0 < \zeta < 1$. In the case of critical damping or overdamping (i.e., $\zeta \geqslant 1$) the form of the equation changes to being a pure exponential with no "ringing." The equations above are valid only for $0 < \zeta < 1$.

By plotting the time domain response of the second order system as a function of ζ, it is possible to deduce a correlation between maximum overshoot, decay time, and peaking in the frequency domain. Figures E-3a through E-3j are calculator printouts of the unit step time domain response $g(t)$ as a function of ζ. Although the resolution is low, it is easy to see the general pattern. As ζ decreases, the transient response becomes more oscillatory. From the closed loop frequency domain plots (Fig. E-1, inverted), it is observed that a low value of ζ produces "peaking" near $\omega = \omega_N$. In general, more peaking in the frequency domain response implies more overshoot and decay time in the time domain response. This correlation is true for higher order systems as well as

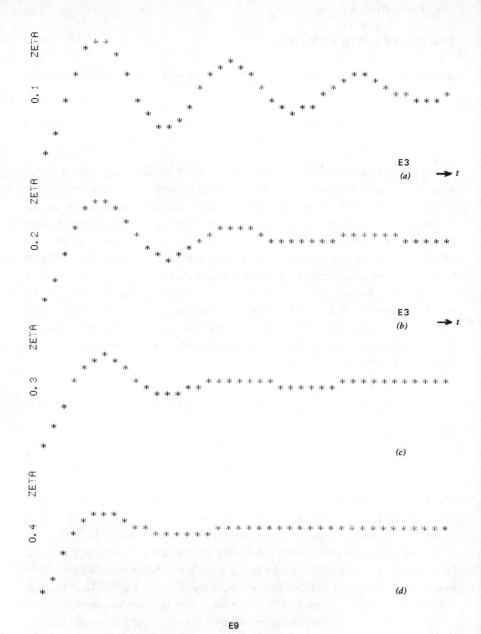

Figure E-3 Calculator printouts of the unit step time domain response $g(t)$ as a function of ζ.

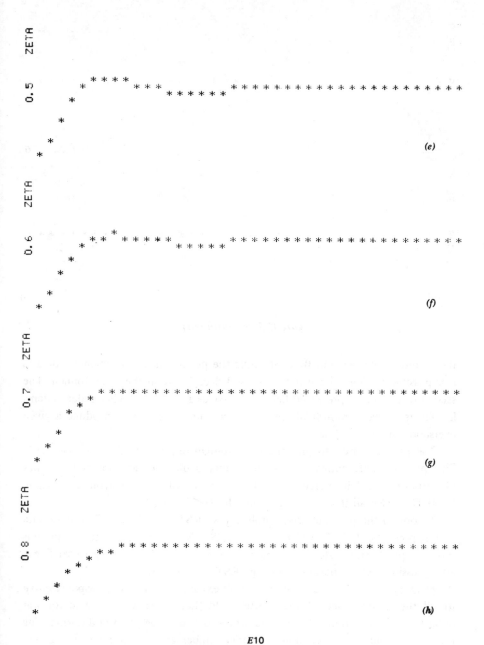

E10

Figure E-3 (*Continued*)

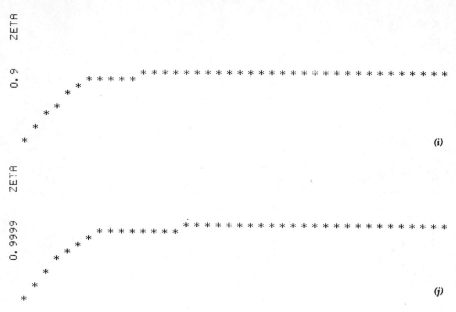

Figure E-3 (*Continued*)

the second order system. By controlling the peaking in the frequency domain, it is possible to control the overshoot and decay time in the time domain. The second order curves presented here can serve as a guide for higher order systems insofar as a given amount of frequency domain peaking will produce a given overshoot and decay time.

The program used to plot the time domain response is given on page 193. This program automatically generates either a plot or data points for values of ζ from 0.1 to 1 in increments of 0.1. The case of $\zeta = 1$ is avoided by using $\zeta = 0.9999$. (Recall that $g(t)$ is valid only for $0 < \zeta < 1$.)

To obtain the plot outputs, simply press "RST" and "RS." To obtain data points, press "RST," "2nd STF 1," and "RS." By setting flag 1, the program is made to print data instead of plotting. Note that it is necessary to set flag 1 after pressing "RST," because pressing "RST" resets all flags.

Memory 14 contains a scale factor that expands the unit step response curve to fill the entire space of the plotter field (i.e., 20 spaces). The factor was obtained by first running data points for $\zeta = 0.1$ (on page 194), and normalizing the largest value of those points to the number 19. The number 19 is used because it is the maximum of the 20 points available (0–19). Since the highest time domain peaking occurs with $\zeta = 0.1$, the greatest value of that data is used. All other curves are scaled by the same factor and are thus in the proper proportion for comparison.

Program listing:

```
000  01   1
001  32   X:T
002  70   RAD
003  01   1          Initialize
004  00   0          ζ register
005  42   STO
006  03   03
007  76   LBL
008  23   LNX
009  53   (
010  01   1
011  01   1
012  75   -
013  43   RCL        Calculate ζ
014  03   03
015  54   )
016  55   ÷
017  01   1
018  00   0
019  95   =
020  42   STO
021  10   10
022  22   INV
023  67   EQ
024  42   STO
025  93   .          If ζ = 1
026  09   9          use
027  09   9          ζ = 0.9999
028  09   9
029  09   9
030  42   STO
031  10   10
032  76   LBL
033  42   STO
034  43   RCL
035  15   15
036  69   OP         Print ZETA
037  04   04         and label
038  43   RCL
039  10   10
040  69   OP
041  06   06
042  04   4          Initialize
043  00   0          ωt register
044  42   STO
045  02   02
046  43   RCL
047  10   10
048  33   X²
049  94   +/-        Calculate
050  85   +          √1-ζ²
051  01   1
052  95   =
053  34   √X
054  42   STO
055  11   11
056  76   LBL
057  33   X²
```

```
058  53   (
059  04   4
060  01   1
061  75   -          Determine
062  43   RCL        ωt
063  02   02
064  54   )
065  55   ÷
066  02   2
067  95   =
068  42   STO
069  00   00
070  07   IFF        Select
071  01   01         graph
072  52   EE         or data
073  71   SBR
074  30   TAN
075  71   SBR
076  38   SIN
077  43   RCL
078  14   14
079  65   ×
080  43   RCL        Plot
081  13   13         graph
082  95   =
083  58   FIX
084  00   00
085  69   OP
086  07   07
087  22   INV
088  58   FIX
089  76   LBL
090  45   YX
091  97   DSZ
092  02   02
093  33   X²
094  98   ADV
095  98   ADV        Loop
096  97   DSZ        control
097  03   03
098  23   LNX
099  98   ADV
100  98   ADV
101  91   R/S        END
102  81   RST
103  76   LBL
104  52   EE
105  71   SBR
106  30   TAN
107  71   SBR        Write
108  38   SIN        data
109  43   RCL
110  00   00
111  99   PRT
112  43   RCL
113  13   13
114  99   PRT
```

E12

```
115  61   GTO
116  45   YX
117  76   LBL
118  30   TAN
119  53   (
120  43   RCL
121  11   11
122  55   ÷
123  43   RCL        Calculate
124  10   10         ψ
125  94   +/-
126  54   )
127  22   INV
128  30   TAN
129  42   STO
130  12   12
131  92   RTN
132  76   LBL
133  38   SIN
134  53   (
135  53   (
136  43   RCL
137  00   00
138  65   ×
139  43   RCL
140  11   11
141  75   -
142  43   RCL
143  12   12
144  54   )
145  38   SIN
146  65   ×
147  43   RCL        Calculate
148  11   11         f(t)
149  35   1/X
150  65   ×
151  53   (
152  43   RCL
153  10   10
154  65   ×
155  43   RCL
156  00   00
157  94   +/-
158  54   )
159  22   INV
160  23   LNX
161  94   +/-
162  85   +
163  01   1
164  54   )
165  42   STO
166  13   13
167  92   RTN
```

Program to obtain g(t)

Memory listing

```
1.              INDEX ωt            00
1.              INDEX ζ             01
39.             INDEX ζ             02
10.                                 03
0.                                  04
0.                                  05
0.                                  06
0.                                  07
0.                                  08
0.                                  09
0.1             ζ²                  10
.9949874371     √1-ζ²               11
-1.470628906    (G)                 12
.1184535973     ψ factor            13
11.04564325     ZETA factor         14
46173713.       "ZETA"              15
```

```
            0.1      ZETA           10.5
            0.5                1.212388824
     .1184535973                      11.
              1.                1.050285276
     .4310281091                      11.5
            1.5                 .8918918535
     .8464231254                      12.
              2.                .7737598914
     1.258070263                      12.5
            2.5                 .7195818086
     1.570415689                      13.
              3.                .7358946489
     1.720135221                      13.5
            3.5                 .8121895528
     1.687821307                      14.
              4.                .9249823694
     1.498325602                      14.5
            4.5                 1.044520039
     1.210756916                      15.
              5.                1.142309381
     .9014493324                      15.5
            5.5                 1.19763589
     .6445377745                      16.
              6.                1.201649802
     .4948944407                      16.5
            6.5                 1.158298949
     .4771771104                      17.
              7.                1.082191085
     .5829593607                      17.5
            7.5                 .9941759757
     0.77584223 4                     18.
              8.                0.915893563
     1.002596843                      18.5
            8.5                 .8646568342
     1.207167854                      19.
              9.                .8498298006
     1.344002905                      19.5
            9.5                 .8714022017
     1.387666453                      20.
            10.                 .9208839764
     1.336851681
```

Data Points of $g(t)$ for $\zeta = 0.1$.

F

Motor Data Sheet and Photographs

Data Sheet and Photographs used with permission from PMI Motors Division, Kollmorgen Corp., Syosset, N.Y.

PMI ENGINEERING DATA

Type U12M4H

All Nominals at 25°c Ambient Except Where Otherwise Stated.

1.0 MOTOR RATINGS

1.1 Continuous Torque @ Rated Speed: 142 Oz.-In.
1.2 Pulse Torque (50ms @ 1% Duty Cycle): 2280 Oz.-In.
1.3 Rated Speed: 2660 RPM
1.4 Rated Voltage[1] : 60 VDC
1.5 Power Out @ Rated Speed: 275 Watts
1.6 Rated Current: 5.9 Amperes
1.7 Maximum Continuous Stall Current: 8.0 Amperes
1.8 Terminal Resistance: 0.75 Ohms

2.0 MOTOR CONSTANTS

2.1 Torque Constant (K_t): 27 Oz.-In./Ampere
2.2 Emf Constant (K_t): 20 Volts/1000 RPM
2.3 Damping Constant (K_b): 4.7 Oz.-In./1000 RPM
2.4 Total Inertia (J_m): 0.023 Oz.-In. Sec.^{2}
2.5 Regulation @ Constant Voltage (R_m)[2]: 2.2 RPM/Oz.-In.
2.6 Armature Inductance (L_a): $<100\mu$ Henries
2.7 Average Friction Torque (T_f): 6.0 Oz.-In.
2.8 Mechanical Time Constant[2]: 0.0053 Sec.
2.9 Power Rate[3]: 1584 KW/Sec.

3.0 THERMAL RESISTANCE

3.1   Uncooled
3.1.1 Armature-To-Case (θ_{AC}): 0.97°C/Watt
3.1.2 Case-To-Ambient (θ_{CA})
3.1.2.1 With 8x16x⅜ Alum. Heat Sink: 0.82°C/Watt
3.1.2.2 With 14x14x⅜ Alum. Heat Sink: 0.66°C/Watt

3.2   Forced Cooling
3.2.1 Armature-To-Ambient (θ_t)
3.2.1.1 With Mass Air Flow of 0.4 lbs./min.: 0.80° C/Watt
3.2.1.2 With Mass Air Flow of 0.8 lbs./min.: 0.51° C/Watt
3.2.1.3 With Mass Air Flow of 2.0 lbs./min.: 0.28° C/Watt

4.0 WEIGHT: 14 lbs.

NOTES:

1. Motor is tested at this voltage for convenience. Other voltages may be used provided maximum armature dissipation is not exceeded. ($P_{MAX} = P_{IN} - P_{OUT} = $ Constant).

2. The speed-torque curve is obtained by using the maximum terminal resistance of the motor at 150°C armature temperature. (Worst condition.)

3. Calculated from the formula,

$$7.01 \times 10^{-3} \times \frac{(\text{Pulse Torque})^2}{\text{Inertia}}$$

Courtesy PMI Motors

GENERAL

1. Maximum allowable armature dissipation,

$$P_{MAX} = \frac{150°C - T_{AMBIENT} (°C)}{\theta_{AC} + \theta_{CA}}$$

2. The curves for forced cooling operation were obtained by modifying the mechanical configuration of the motor to accept the required air flow. These motors are available on special request.
   The maximum allowable armature dissipation in this case is calculated as follows:

$$P_{MAX} = \frac{150°C - T_{AMBIENT} (°C)}{\theta_t}$$

3. Mass Air Flow (lbs./min.) = Air Volume (cfm) x Density (lbs./ft.^{3})

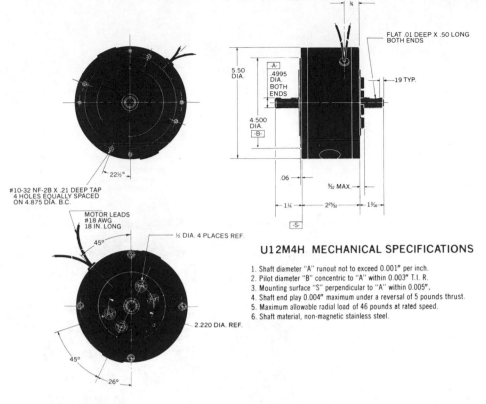

#10-32 NF-2B X .21 DEEP TAP
4 HOLES EQUALLY SPACED
ON 4.875 DIA. B.C.

22½°

5.50 DIA.

-A-
.4995 DIA. BOTH ENDS

4.500 DIA.
-B-

FLAT .01 DEEP X .50 LONG BOTH ENDS

¾

.19 TYP.

.06

⁹⁄₃₂ MAX.

1¼   2²⁵⁄₃₂   1³⁄₁₆

-S-

MOTOR LEADS
#18 AWG
18 IN. LONG

45°

½ DIA. 4 PLACES REF.

2.220 DIA. REF.

45°

26°

U12M4H MECHANICAL SPECIFICATIONS

1. Shaft diameter "A" runout not to exceed 0.001" per inch.
2. Pilot diameter "B" concentric to "A" within 0.003" T.I. R.
3. Mounting surface "S" perpendicular to "A" within 0.005".
4. Shaft end play 0.004" maximum under a reversal of 5 pounds thrust.
5. Maximum allowable radial load of 46 pounds at rated speed.
6. Shaft material, non-magnetic stainless steel.

AVERAGE PERFORMANCE CHARACTERISTICS

LIMIT OF ALLOWANCE
CONTINUOUS OPERATION
(UNCOOLED)

The run current at any operating condition is obtained as follows:

$$I_{RUN} = \frac{K_D \times \frac{N}{1000} + T_f + T_1}{K_T}$$

Courtesy PMI Motors

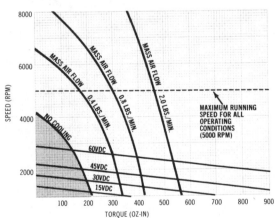

MASS AIR FLOW

MASS AIR FLOW

MASS AIR FLOW

0.4 LBS./MIN.

0.8 LBS./MIN.

2.0 LBS./MIN.

MAXIMUM RUNNING
SPEED FOR ALL
OPERATING
CONDITIONS
(5000 RPM)

NO COOLING

60VDC

45VDC

30VDC

15VDC

SPEED (RPM)

TORQUE (OZ-IN)

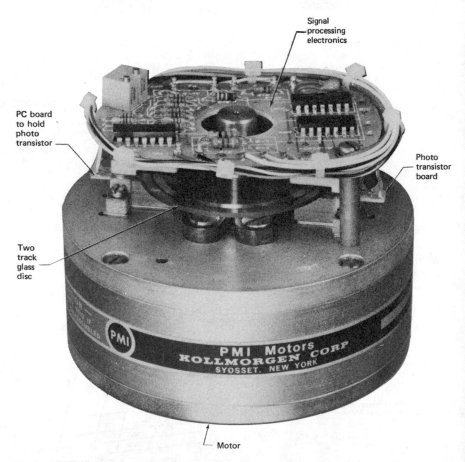

Signal
processing
electronics

PC board
to hold
photo
transistor

Photo
transistor
board

Two
track
glass
disc

PMI

PMI Motors
KOLLMORGEN CORP
SYOSSET, NEW YORK

Motor

Courtesy PMI Motors

198

Phototransistor
board

Courtesy PMI Motors

199

References

BOOKS

DelToro, Vincent and Parker, Sydney R. *Principles of Control Systems Engineering.* New York: McGraw-Hill, Inc., 1960.

DiStefano, Joseph, Stubberud, Allen, and Williams, Ivan. *Theory and Problems of Feedback and Control Systems.* New York: Schaum Publishing Co., 1967.

Electro-Craft Corp., Ed. *DC Motors, Speed Controls, Servo Systems.* Minnesota: Electro-Craft Corp., 1978.

Elgerd, Olle I. *Control Systems Theory.* New York: McGraw-Hill, Inc., 1967.

Eveleigh, Virgil W. *Introduction to Control Systems Design.* New York: McGraw-Hill, Inc., 1972.

Gardner, Floyd M. *Phaselock Techniques.* New York: John Wiley & Sons, Inc., 1966.

Gardner, Floyd M. *Phaselock Techniques.* New York: John Wiley & Sons, Inc., 1979.

Texas Instruments Learning Center. *Personal Programming: A Complete Owner's Manual for TI Programmable 58/59.* Texas: Texas Instruments Inc., 1977.

Texas Instruments Learning Center. *Master Library: TI Programmable 58/59.* Texas: Texas Instruments Inc., 1977.

ARTICLES

Geiger, Dana F. "Velocity Control of DC Motors by Use of Phaselock Servo Techniques." Application Note. New York: Kollmorgen Corporation, 1973.

Geiger, Dana F. "Velocity Control for PMI Motors." Application Note. New York: Kollmorgen Corporation, 1978.

Nash, Garth. "Phase Locked Loop Design Fundamentals." Application Note #535. Arizona: Motorola Semiconductor Products, Inc.

RCA COS/MOS Integrated Circuits. "The RCA COS/MOS Phase Locked Loop" (pp. 612–615). Application Note. New Jersey: RCA Corporation, 1977.

Signetics Corp. "Phase Locked Loops" (pp. 807–860). *Signetics Analog Data Manual.* California: Signetics Corp., 1977.

Index